Estimation of Cortical Connectivity in Humans: Advanced Signal Processing Techniques

Estimation of Cortical Connectivity in Humans: Advanced Signal Processing Techniques
Laura Astolfi and Fabio Babiloni

ISBN: 978-3-031-00494-0   paperback
ISBN: 978-3-031-01622-6   ebook

DOI 10.1007/978-3-031-01622-6

A Publication in the Springer series

SYNTHESIS LECTURES ON BIOMEDICAL ENGINEERING #13

Lecture #13
Series Editor : John D. Enderle, University of Connecticut

Series ISSN

ISSN 1930-0328   print
ISSN 1930-0336   electronic

Estimation of Cortical Connectivity in Humans: Advanced Signal Processing Techniques

Laura Astolfi and Fabio Babiloni
Università degli Studi di Roma La Sapienza Italy

SYNTHESIS LECTURES ON BIOMEDICAL ENGINEERING #13

ABSTRACT

In the last ten years many different brain imaging devices have conveyed a lot of information about the brain functioning in different experimental conditions. In every case, the biomedical engineers, together with mathematicians, physicists and physicians are called to elaborate the signals related to the brain activity in order to extract meaningful and robust information to correlate with the external behaviour of the subjects. In such attempt, different signal processing tools used in telecommunications and other field of engineering or even social sciences have been adapted and re-used in the neuroscience field. The present book would like to offer a short presentation of several methods for the estimation of the cortical connectivity of the human brain. The methods here presented are relatively simply to implement, robust and can return valuable information about the causality of the activation of the different cortical areas in humans using non invasive electroencephalographic recordings. The knowledge of such signal processing tools will enrich the arsenal of the computational methods that a engineer or a mathematician could apply in the processing of brain signals.

KEYWORDS

Multivariate autoregressive model (MVAR), High resolution EEG, Directed Transfer Function (DTF), Partial Directed Coherence (PDC), Realistic head modeling, cortical imaging

Contents

Foreword

This book describes some advanced mathematical signal processing techniques applied to the estimation of the cortical connectivity in humans from non invasive electroencephalographic recordings. Although it may be thought that mathematics could not be the proper tool for a full comprehension of the brain functions, this is often not the case. In the last 10 years many different brain-imaging devices have conveyed a lot of information about the brain functioning in different experimental conditions. In every case, the biomedical engineers, together with mathematicians, physicists, and physicians are called in to elaborate the signals related to the brain activity in order to extract meaningful and robust information to correlate with the external behavior of the subjects. In such attempts, different signal processing tools used in telecommunications and other fields of engineering or even social sciences have been adapted and reused in the neuroscience field.

The present book offers a short presentation of several methods for the estimation of the cortical connectivity of the human brain. The methods presented here are relatively simple to implement, robust, and can return valuable information about the causality of the activation of the different cortical areas in humans using noninvasive electroencephalographic recordings. For these reasons, they are among the most used tools in the modern neuroscientific research. The knowledge of such signal processing tools will enrich the arsenal of the computational methods that an engineer or a mathematician could apply in the processing of brain signals. In our understanding, a comprehensive presentation of these connectivity estimation methods is lacking in literature and this book could represent a good step in filling the void. With this book we would like to be in agreement with a citation of the famous physicist Ludwig Boltzmann, "There is nothing more practical than a good theory". We think that this is still true a hundred years after him.

Laura Astolfi
Fabio Babiloni

INTRODUCTION

In recent years, great advancements have been made in understanding the mechanisms of the functioning of the human brain. Technological developments such as functional magnetic resonance imaging (fMRI), positron emission tomography (PET), and magnetoencephalography (MEG) have made possible the mapping of the images of cerebral activity from hemodynamic, metabolic or electromagnetic measurements. Among these brain imaging techniques, electroencephalography (EEG) is unique in terms of simplicity, accessibility, and temporal resolution, and has been viewed with renewed interest in recent years, thanks to the use of advanced methods of analysis and interpretation of its data. These methods are able to improve the spatial resolution of conventional EEG, making it possible to address the analysis of the brain activity in a noninvasive way using the temporal resolution of brain phenomena (of the order of milliseconds). With high-resolution EEG, it is now possible to obtain cortical activation maps describing the activity of the brain at the cortical level during the execution of a given experimental task.

Simple imaging of regions of the brain activated during particular tasks does not, however, convey the information about how these regions communicate with each other for making the execution of the task possible. The concept of brain connectivity is viewed as central to the understanding of the organized behavior of cortical regions, beyond the simple mapping of their activities [1,2]. Such behavior is thought to be based on the interaction between different cortical sites and differently specialized ones. Cortical connectivity estimation aims to describe these interactions in connectivity patterns, which hold the direction and strength of the information flow between cortical areas. To this purpose, several methods have been developed and applied to data gathered from hemodynamic and electromagnetic techniques [3–7]. Two main definitions of brain connectivity have been proposed during recent years: *functional* and *effective* connectivity [8]. Functional connectivity is defined as the temporal correlation between spatially remote neurophysiologic events. Effective connectivity is defined as the simplest brain circuit which would produce the same temporal relationship between cortical sites as observed experimentally.

As for the functional connectivity, the methods proposed in literature typically involve estimation of some covariance properties between different time series. These properties are measured from different spatial sites during motor and cognitive tasks by EEG and fMRI techniques [4,5,7,9].

Structural equation modeling (SEM) is a technique that has been used recently to assess the connectivity between cortical areas in humans from hemodynamic and metabolic measurements [3,10–12]. The basic idea of SEM considers the covariance structure of the data [10]. The estimation of effective cortical connectivity obtained from fMRI data has, however, a low temporal resolution (of the order of seconds) which is far from the time scale in which the brain normally operates. Hence, it is interesting to know whether the SEM technique can be applied to cortical activity which are obtained by applying linear inverse techniques to high-resolution EEG data [5,13–15].

As important information in the EEG signals are coded in frequency domain rather than time domain (reviewed in [16]), attention was focused on detecting frequency-specific interactions in EEG or MEG signals; for instance, the coherence between the activity of pairs of channels [17–19]. However, coherence analysis does not have a directional nature (i.e., it just examines whether a link exists between two neural structures by describing instances when they are in synchronous activity) and does not provide directly the direction of the information flow. In this respect, multivariate spectral techniques such as directed transfer function (DTF) or partial directed coherence (PDC) were proposed [20,21] for determining the directional influences between any given pair of channels in a multivariate data set. Both DTF and PDC [21,22] rely on the key concept of Granger causality between time series [23], according to which an observed time series $x(n)$ causes another series $y(n)$ if the knowledge of $x(n)$'s past significantly improves the prediction of $y(n)$; this relation between time series is not reciprocal, i.e., $x(n)$ may cause $y(n)$ without $y(n)$ necessarily causing $x(n)$. This lack of reciprocity allows the evaluation of the direction of information flow between structures. These estimators are able to characterize at the same time both the directional and spectral properties of the brain signals, and they require only one multivariate autoregressive (MVAR) model estimated from all the EEG channels. The advantages of MVAR modeling of multichannel EEG signals were stressed recently [24] by demonstrating the advantages of multivariate methods with respect to the pairwise autoregressive approach, in terms of both accuracy and computational cost. In order to fully characterize the techniques presented we test them on simulated EEG data whose connectivity characteristics are known in advance, and then we finally apply such methods to human data obtained from high resolution EEG recordings. As a novelty, the application of all the proposed methodologies (SEM, DTF, and PDC) was performed using the cortical signals estimated from high-resolution EEG recordings which exhibit a higher spatial resolution than conventional cerebral electromagnetic measurements. To correctly estimate the cortical signals we used multicompartment head models (scalp, skull, dura mater, and cortex) constructed from individual MRI, a distributed source model, and a regularized linear inverse source estimates of cortical current density.

Chapters I–III discuss the simulation studies in which different main factors (signal-to-noise ratio, cortical activity duration, frequency band, etc.) are systematically imposed in the generation of test signals, and the errors in the estimated connectivity are evaluated by analysis of variance (ANOVA). In particular, we first explore the behavior of the most advanced estimators of effective and functional connectivity – SEM, DTF, dDTF, and PDC – in a simulation context and under different practical conditions.

For SEM, which involves the definition of an a priori connectivity model, the simulation study is designed to answer the following questions:

1. What is the influence of a variable signal-to-noise ratio (SNR) level on the accuracy of the pattern connectivity estimation obtained by SEM?

2. What is the amount of data necessary to get good accuracy of the estimation of connectivity between cortical areas?

3. How are the SEM performances degraded by an imprecise anatomical model formulation? Is it able to perform a good estimation of connectivity pattern when connections between the cortical areas are not correctly assumed? Which kind of errors should be avoided?

For the three multivariate estimators of functional connectivity – DTF, dDTF, and PDC – the experimental design focused on the following questions:

1. How are the connectivity pattern estimators influenced by different factors affecting the EEG recordings such as the signal-to-noise ratio and the amount of data available?

2. How do the estimators discriminate between the direct or indirect causality patterns?

3. What is the most effective method for estimating a connectivity model under the conditions usually encountered in standard EEG recordings?

These questions are addressed via simulations using predefined connectivity schemes linking several cortical areas. The estimation process retrieves the cortical connections between the areas under different experimental conditions. The connectivity patterns estimated by the four techniques are compared with those imposed on the simulated signals, and different error measures are computed and subjected to statistical multivariate analysis. The statistical analysis showed that during the simulations, SEM, DTF, and PDC estimators are able to estimate the imposed connectivity patterns under reasonable operating conditions. It was possible to conclude that the estimation of cortical connectivity can be performed not only with hemodynamic measurements, but also with EEG signals obtained from advanced computational techniques.

After giving a full description of the properties of these connectivity estimators for high resolution EEG recordings, the results of their application to human data relating to different experimental tasks such as finger tapping, Stroop test, and movement imagination are discussed. After the simulation tests, SEM, DTF, and PDC are applied to different sets of experimental data relating to motor and cognitive tasks (Chapters IV–VI). The motor task examined is a fast repetitive finger tapping, while the cognitive task involved recordings during the Stroop test, often employed in studies of selective attention, and found to be sensitive to prefrontal damage. The data employed are cortical estimates obtained from high-resolution EEG recordings using very advanced techniques which exhibit a higher spatial resolution than conventional cerebral electromagnetic measurements. We also briefly describe the high-resolution EEG techniques, including the use of a large number of scalp electrodes, which are realistic models of the head derived from structural magnetic resonance images (MRIs), and advanced processing methodologies related to solutions of linear inverse problems. The results of the estimation of the effective and functional connectivity from data recorded during finger tapping test, Stroop test and movement imagination test are presented.

One of the possible problems in the approach presented above is the hytpothesis that the gathered EEG data are stationary. However, this property of the data cannot be easily assumed. In fact, under this hypothesis, the methodology proposed for the DTF and PDC techniques is valid. Then the connectivity estimation methods for dealing with nonstationary EEG data are highly desirable. In Chapter VII we propose a methodology for the estimation of cortical connectivity extended to the time–frequency domain, based on the use of adaptive multivariate models. Such an approach allows extention of the connectivity analysis to nonstationary data and monitoring the rapid changes in the connectivity between cortical areas during an experimental task. The performances of the time-varying estimators are tested by means of simulations performed on the basis of a predefined connectivity scheme linking different cortical areas. Cortical connections between the areas are retrieved by the estimation process under different experimental conditions, and the results obtained for the different methods are evaluated by statistical analyses.

Finally, as an example of the results that can be obtained by this technique, the application of the simulation study to real data is proposed in Chapter VIII. For this purpose, we applied the time-varying technique to the cortical activity estimated in particular regions of interest (ROIs) of the cortex, and obtained high-resolution EEG recordings during the execution of a combined foot–lips movement in a group of normal subjects.

The experimental data presented here as practical results of the estimation of cortical connectivity in humans during motor and cognitive tasks were obtained from the Laboratory of High Resolution EEG of the University of Rome "La Sapienza" and from the Laboratory of Neuroelectric Imaging and Brain Computer Interface of the S. Lucia Foundation. In addition, part of the experimental data employed were provided by the Department of Biomedical Engineering, University of Minnesota, Minneapolis, USA, and by the Department of Psychology and Beckman Institute Biomedical Imaging Center, University of Illinois at Urbana-Champaign, Illinois, USA, in the framework of a scientific cooperation with the University of Rome.

REFERENCES

[1] L. Lee, L. M. Harrison, and A. Mechelli, "The functional brain connectivity workshop: Report and commentary," *NeuroImage*, vol. 19, pp. 457–465, 2003.

[2] B. Horwitz, "The elusive concept of brain connectivity," *NeuroImage*, vol. 19, pp. 466–470, 2003, doi:10.1016/S1053-8119(03)00112-5.

[3] C. Buchel and K. J. Friston, "Modulation of connectivity in visual pathways by attention: Cortical interactions evaluated with structural equation modeling and fMRI," *Cereb. Cortex*, vol. 7, no. 8, pp. 768–778, 1997, doi:10.1093/cercor/7.8.768.

[4] A. Urbano, C. Babiloni, P. Onorati, and F. Babiloni, "Dynamic functional coupling of high resolution EEG potentials related to unilateral internally triggered one-digit movements," *Electroencephalogr. Clin. Neurophysiol.*, vol. 106, no. 6, pp. 477–487, 1998, doi:10.1016/S0013-4694(97)00150-8.

[5] A. S. Gevins, B. A. Cutillo, S. L. Bressler, N. H. Morgan, R. M. White, J. Illes, and D. S. Greer, "Event-related covariances during a bimanual visuomotor task. II. Preparation and feedback," *Electroencephalogr. Clin. Neurophysiol.*, vol. 74, pp. 147–160, 1989, doi:10.1016/0168-5597(89)90020-8.

[6] M. Taniguchi, A. Kato, N. Fujita, M. Hirata, H. Tanaka, T. Kihara, H. Ninomiya, N. Hirabuki, H. Nakamura, S. E. Robinson, D. Cheyne, and T. Yoshimine, "Movement-related desynchronization of the cerebral cortex studied with spatially filtered magnetoencephalography," *NeuroImage*, vol. 12, no. 3, pp. 298–306, 2000, doi:10.1006/nimg.2000.0611.

[7] A. Brovelli, M. Ding, A. Ledberg, Y. Chen, R. Nakamura, and S. L. Bressler, "Beta oscillations in a large-scale sensorimotor cortical network: Directional influences revealed by Granger causality," *Proc. Natl. Acad. Sci. U.S.A.*, vol. 101, no. 26, pp. 9849–9854, Jun. 29, 2004, doi:10.1073/pnas.0308538101.

[8] K. J. Friston, "Functional and effective connectivity in neuroimaging: A synthesis," *Human Brain Mapp.*, vol. 2, pp. 56–78, 1994, doi:10.1002/hbm.460020107.

[9] L. Jancke, R. Loose, K. Lutz, K. Specht, and N. J. Shah, "Cortical activations during paced finger-tapping applying visual and auditory pacing stimuli," *Brain Res. Cogn. Brain Res.*, vol. 10, no. 1–2, pp. 51–66, 2000, doi:10.1016/S0926-6410(00)00022-7.

[10] K. A. Bollen, *Structural Equations with Latent Variables*. New York: Wiley; 1989.

[11] A. R. McIntosh and F. Gonzalez-Lima, "Structural equation modeling and its application to network analysis in functional brain imaging," *Hum. Brain Mapp.*, vol. 2, pp. 2–22, 1994, doi:10.1002/hbm.460020104.

[12]   R. Schlosser, T. Gesierich, B. Kaufmann, G. Vucurevic, S. Hunsche, J. Gawehn, and P. Stoeter, "Altered effective connectivity during working memory performance in schizophrenia: A study with f MRI and structural equation modeling," *NeuroImage*, vol. 19, no. 3, pp. 751–763, 2003, doi:10.1016/S1053-8119(03)00106-X.

[13]   P. L. Nunez, *Neocortical Dynamics and Human EEG Rhythms*. New York: Oxford University Press, 1995.

[14]   F. Babiloni, C. Babiloni, L. Locche, F. Cincotti, P. M. Rossini, and F. Carducci, "High-resolution electroencephalogram: Source estimates of Laplacian-transformed somatosensory-evoked potentials using a realistic subject head model constructed from magnetic resonance images," *Med. Biol. Eng. Comput.*, vol. 38, no. 5, pp. 512–519, Sep. 2000, doi:10.1007/BF02345746.

[15]   F. Babiloni, C. Babiloni, F. Carducci, G. L. Romani, P. M. Rossini, L. M. Angelone, and F. Cincotti, "Multimodal integration of high-resolution EEG and functional magnetic resonance imaging data: A simulation study," *NeuroImage*, vol. 19, no. 1, pp. 1–15, May 2003.

[16]   G. Pfurtscheller and F. H. Lopes da Silva, "Event-related EEG/MEG synchronization and desynchronization: Basic principles," *Clin. Neurophysiol.*, vol. 110, no. 11, pp. 1842–1857, Nov. 1999, doi:10.1016/S1388-2457(99)00141-8.

[17]   S. L. Bressler, "Large-scale cortical networks and cognition," *Brain Res. Brain Res. Rev.*, vol. 20, no. 3, pp. 288–304, 1995, doi:10.1016/0165-0173(94)00016-I.

[18]   J. Gross, J. Kujala, M. Hämäläinen, L. Timmermann, A. Schnitzler, and R. Salmelin, "Dynamic imaging of coherent sources: Studying neural interactions in the human brain," *Proc. Natl. Acad. Sci. U.S.A.*, vol. 98, no. 2, pp. 694–699, 2001, doi:10.1073/pnas.98.2.694.

[19]   J. Gross, L. Timmermann, J. Kujala, R. Salmelin, and A. Schnitzler, "Properties of MEG tomographic maps obtained with spatial filtering," *NeuroImage*, vol. 19, pp. 1329–1336, 2003, doi:10.1016/S1053-8119(03)00101-0.

[20]   M. Kaminski and K. Blinowska, "A new method of the description of the information flow in the brain structures," *Biol. Cybern.*, vol. 65, pp. 203–210, 1991, doi:10.1007/BF00198091.

[21]   L. A. Baccalà and K. Sameshima, "Partial directed coherence: A new concept in neural structure determination," *Biol. Cybern.*, vol. 84, pp. 463–474, 2001, doi:10.1007/PL00007990.

[22]   M. Kaminski, M. Ding, W. A. Truccolo, and S. Bressler, "Evaluating causal relations in neural systems: Granger causality, directed transfer function and statistical assessment of significance," *Biol. Cybern.*, vol. 85, pp. 145–157, 2001, doi:10.1007/s004220000235.

[23]   C. W. J. Granger, "Investigating causal relations by econometric models and cross-spectral methods," *Econometrica*, vol. 37, pp. 424–438, 1969, doi:10.2307/1912791.

[24]   R. Kus, M. Kaminski, and K. J. Blinowska, "Determination of EEG activity propagation: Pairwise versus multichannel estimate," *IEEE Trans. Biomed. Eng.*, vol. 51, no. 9, pp. 1501–1510, Sep. 2004, doi:10.1109/TBME.2004.827929.

CHAPTER 1

Estimation of the Effective Connectivity from Stationary Data by Structural Equation Modeling

1.1 STRUCTURAL EQUATION MODELING

In structural equation model (SEM), the parameters are estimated by minimizing the difference between the observed covariances and those implied by a structural or path model [1,2]. In terms of neural systems, a measure of covariance represents the degree to which the activities of two or more regions are related.

The SEM consists of a set of linear structural equations, containing observed variables and parameters defining causal relationships among the variables. Variables in the equation system can be endogenous (i.e., dependent on other variables in the model) or exogenous (independent of the model itself). The SEM specifies the causal relationship among the variables, describes the causal effects, and assigns the explained and the unexplained variances.

Let us consider a set of variables (expressed as deviations from their means) with N observations. In this study, these variables represent the activity estimated in each cortical region of the brain, obtained with the procedures described in the chapters that follow.

The SEM for these variables is the following:

$$\mathbf{y} = \mathbf{By} + \mathbf{\Gamma x} + \mathbf{\zeta} \qquad (1.1)$$

where

$\mathbf{y}$ is an $m \times 1$ vector of dependent (endogenous) variables;

$\mathbf{x}$ is an $n \times 1$ vector of independent (exogenous) variables;

$\mathbf{\zeta}$ is an $m \times 1$ vector of equation errors (random disturbances);

$\mathbf{B}$ is an $m \times m$ matrix of coefficients of the endogenous variables;

$\mathbf{\Gamma}$ is an $m \times n$ matrix of coefficients of the exogenous variables;

$\mathbf{\zeta}$ is assumed to be uncorrelated with the exogenous variables;

$\mathbf{B}$ is supposed to have zeros in its diagonal (i.e., an endogenous variable does not influence itself) and to satisfies the assumption that $(\mathbf{I} - \mathbf{B})$ is nonsingular, where $\mathbf{I}$ is the identity matrix.

The covariance matrices of this model are the following:

$\mathbf{\Phi} = E[\mathbf{x}\mathbf{x}^T]$ is an $n \times n$ covariance matrix of the exogenous variables;

$\mathbf{\Psi} = E[\mathbf{\zeta}\mathbf{\zeta}^T]$ is an $m \times m$ covariance matrix of the errors.

If $\mathbf{z}$ is a vector containing all the $p = m + n$ variables, exogenous and endogenous, in the following order

$$\mathbf{z}^T = \begin{bmatrix} x_1 \cdots x_n & y_1 \cdots y_m \end{bmatrix} \tag{1.2}$$

the observed covariances can be expressed as

$$\mathbf{\Sigma}_{\text{obs}} = (1/(N-1))\mathbf{z}\mathbf{z}^T \tag{1.3}$$

where $\mathbf{z}$ is the $p \times N$ matrix of the p observed variables for N observations.

The covariance matrix implied by the model can be obtained as follows:

$$\mathbf{\Sigma}_{\text{mod}} = E[\mathbf{z}^T\mathbf{z}] = \begin{bmatrix} E[\mathbf{x}\mathbf{x}^T] & E[\mathbf{x}\mathbf{y}^T] \\ E[\mathbf{y}\mathbf{x}^T] & E[\mathbf{y}\mathbf{y}^T] \end{bmatrix} \tag{1.4}$$

where

$$E[\mathbf{y}\mathbf{y}^T] = E[(\mathbf{I}-\mathbf{B})^{-1}(\mathbf{\Gamma}\mathbf{x}+\zeta)(\mathbf{\Gamma}\mathbf{x}+\zeta)^T((\mathbf{I}-\mathbf{B})^{-1})^T]$$
$$= (\mathbf{I}-\mathbf{B})^{-1}(\mathbf{\Gamma}\mathbf{\Phi}\mathbf{\Gamma}+\mathbf{\Psi})((\mathbf{I}-\mathbf{B})^{-1})^T \tag{1.5}$$

As the errors ζ are not correlated with $\mathbf{x}$

$$E[\mathbf{x}\mathbf{x}^T] = \mathbf{\Phi} \tag{1.6}$$

$$E[\mathbf{x}\mathbf{y}^T] = (\mathbf{I}-\mathbf{B})^{-1}\mathbf{\Phi} \tag{1.7}$$

$$E[\mathbf{y}\mathbf{x}^T] = ((\mathbf{I}-\mathbf{B})^{-1}\mathbf{\Phi})^T \tag{1.8}$$

and as $\mathbf{\Sigma}_{\text{mod}}$ is symmetric. The resulting covariance matrix, in terms of the model parameters, is given by the following:

$$\mathbf{\Sigma}_{\text{mod}} = \begin{bmatrix} \mathbf{\Phi} & (\mathbf{I}-\mathbf{B})^{-1}\mathbf{\Phi} \\ ((\mathbf{I}-\mathbf{B})^{-1}\mathbf{\Phi})^T & (\mathbf{I}-\mathbf{B})^{-1}(\mathbf{\Gamma}\mathbf{\Phi}\mathbf{\Gamma}+\mathbf{\Psi})((\mathbf{I}-\mathbf{B})^{-1})^T \end{bmatrix} \tag{1.9}$$

Without other constraints, the problem of minimizing the differences between the observed covariances and those implied by the model is not determined, because the number of variables (elements of matrices $\mathbf{B}, \mathbf{\Gamma}, \mathbf{\Psi}$, and $\mathbf{\Phi}$) is greater than the number of equations $(m+n)(m+n+1)/2$. For this reason, the SEM technique is based on *a priori* formulation of a model depending on the anatomic and physiologic constraints. This model implies the existence of just some causal relationships among variables represented by arcs in a "path" diagram; all parameters related to arcs that are not present in the hypothesized model are set to zero. For this reason, all the parameters to be estimated are called free parameters. If t is the number of free parameters, then $t \leq (m+n)(m+n+1)/2$.

These parameters are estimated by minimizing a function of the observed and implied covariance matrices. The most widely used objective function for SEM is the maximum likelihood (ML) function:

$$F_{\mathrm{ML}} = \log |\mathbf{\Sigma}_{\mathrm{mod}}| + \mathrm{tr}(\mathbf{\Sigma}_{\mathrm{obs}}\mathbf{\Sigma}_{\mathrm{mod}}^{-1}) - \log |\mathbf{\Sigma}_{\mathrm{obs}}| - p \qquad (1.10)$$

where $\mathrm{tr}(\cdot)$ is the trace of the matrix. In the context of multivariate normally distributed variables, the minimum of the maximum likelihood function when multiplied by $N-1$, follows a χ^2 distribution with $(p(p+1)/2) - t$ degrees of freedom, where t is the number of parameters to be estimated and p is the total number of observed variables (endogenous + exogenous). The χ^2 statistics test can then be used to infer the statistical significance of the structural equation model obtained. In the present study, the publicly available software LISREL [3] was used for the implementation of the SEM technique.

1.2 SIMULATION STUDY

We adopted an experimental design that analyzes the recovery of the connectivity of an estimated model with respect to an imposed one. This has been built under different levels of the main factors, the signal-to-noise ratio (SNR) and the temporal duration (LENGTH); they have been imposed during the generation of a set of test signals simulating cortical average activations and were obtained by starting from actual cortical data (estimated with the high-resolution EEG procedures available at the High Resolution EEG Laboratory at the University of Rome "La Sapienza").

1.2.1 Signal Generation

Different sets of test signals have been generated in order to fit an imposed connectivity pattern (shown in Figs. 1.1A, 1.2A, and 1.3A) and maintain the imposed levels LENGTH and SNR. In the following, in order to use a more compact notation, the signals have been represented with the $\mathbf{z}$ vector defined in Eq. (1.2), containing both the endogenous and the exogenous variables.

Channel z_1 is a reference source waveform, estimated from a high-resolution EEG (128 electrodes) recording in a healthy subject, during the execution of random self-paced movements of the right finger.

Signals z_2, z_3, and z_4 were obtained by the contribution of signals from all other channels with an amplitude variation and zero mean uncorrelated white noise processes with appropriate variances, as shown in the following:

$$\mathbf{Z}[k] = \mathbf{A} * \mathbf{z}[k] + \mathbf{W}[k] \qquad (1.11)$$

where $\mathbf{z}[k]$ is the 4×1 vector of signals, $\mathbf{W}[k]$ is the 4×1 noise vector and $\mathbf{A}$ is the 4×4 parameters matrix obtained from the $\mathbf{\Gamma}$ and $\mathbf{B}$ matrices as follows:

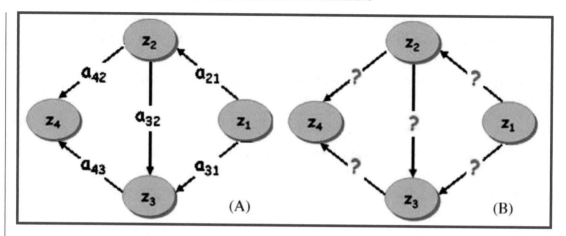

FIGURE 1.1: Correct model. (A) Connectivity patterns imposed in the generation of simulated signals. $z_1, \ldots, z_4$ represent the average activities in the four cortical areas. Values on the arcs represent the strength of the connections ($a_{21} = 1.4; a_{31} = 1.1; a_{32} = 0.5; a_{42} = 0.7; a_{43} = 1.2$). (B) Connectivity model used for the parameter estimation. (Published with permission from [6].)

$$
A = \begin{bmatrix} 0 & 0 & 0 & 0 \\ \gamma_1 & \beta_{11} & \beta_{12} & \beta_{13} \\ \gamma_2 & \beta_{21} & \beta_{22} & \beta_{23} \\ \gamma_3 & \beta_{31} & \beta_{32} & \beta_{33} \end{bmatrix} = \begin{bmatrix} a_{11} & \cdots & & a_{14} \\ \vdots & \ddots & & \vdots \\ & & \ddots & \\ a_{41} & \cdots & & a_{44} \end{bmatrix}
\qquad (1.12)
$$

where β_{ij} stands for the generic (i, j) element of the **B** matrix and γ_i is the ith element of the matrix **Γ**.

All procedures of signal generation were repeated under the following conditions:

SNR factor levels = [1, 3, 5, 10, 100];

LENGTH factor levels (in seconds) = [60, 190, 310, 610]. This corresponds, for instance, to [120, 380, 620, 1220] EEG epochs, each of which is 500 ms long.

It is worth noting that the levels chosen for both SNR and LENGTH factors cover the typical range for the cortical activity estimated with high resolution EEG techniques.

1.2.2   Parameter Estimation

The set of simulated signals generated as described above was given as input to the program LISREL for the estimation of SEM parameters. As mentioned in the section 1.1 of this Chapter, SEM needs a model based on a previous information on the anatomic connections on which the estimate is successively performed. For this reason, its performance has been observed in different situations

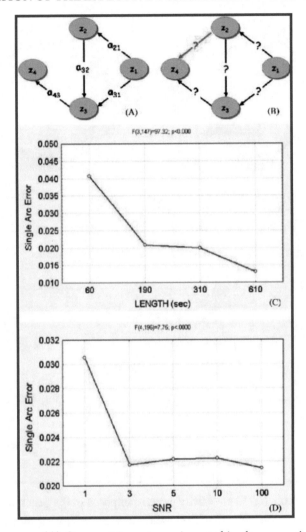

FIGURE 1.2: Arc in excess. (A) Connectivity patterns imposed in the generation of simulated signals. Values on the arcs represent the strength of the connection ($a_{21} = 1.4; a_{31} = 1.1; a_{32} = 0.5; a_{43} = 1.2$). (B) Connectivity model used for the estimation of the parameter. Results of ANOVA performed on the error committed on the arc in excess a_{42} (Single Arc Error). (C) Plot of means with respect to signal LENGTH as a function of time (in seconds). ANOVA shows a high statistical significance of the LENGTH factor ($F = 97.32, p < 0.0001$). Duncan's post hoc test at 5% level of significance shows statistically no significant difference between a signal length of 190 or 310 s (25 or 40 trials, 7.5 s per trial). (D) Plot of means with respect to SNR. A statistical influence of the SNR factor on the error in the evaluation of the presence of arc a_{42} is shown ($F = 7.75, p < 0.0001$). Duncan's post hoc test at 5% level of significance shows no statistically significant differences between levels 3, 5, 10, and 100 of the SNR factor. (Published with permission from [6].)

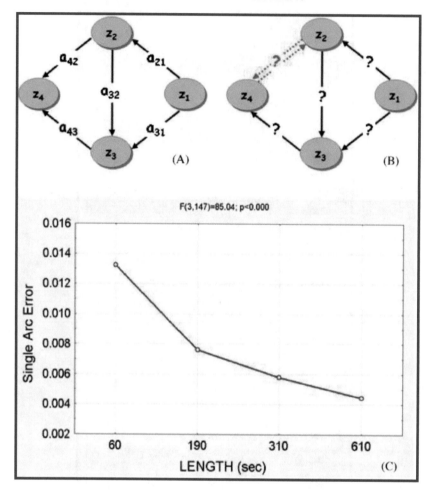

FIGURE 1.3: Arc direction. (A) Connectivity patterns imposed in the generation of simulated signals. Values on the arcs represent the strength of the connections ($a_{21} = 1.4$; $a_{31} = 1.1$; $a_{32} = 0.5$; $a_{42} = 0.7$; $a_{43} = 1.2$). (B) Connectivity model used for the parameter estimation. No assumption has been made on the direction of arc a_{42} (both directions are present in the model). (C) Results of ANOVA performed on the error committed on the wrong direction arc a_{24}, not present in the imposed model (Single Arc Error). Plot of means with respect to signal LENGTH as a function of time (in seconds). ANOVA shows a high statistical significance of the LENGTH factor($F = 85.04, p < 0.0001$). Duncan's post hoc test at 5% level of significance shows statistically significant differences between all levels of LENGTH. (Published with permission from [6].)

where the connections between the four cortical areas are not always correctly assumed. The situations analyzed were as follows:

1. an identical connectivity graph between the generated and the estimated model;

2. a different number of connectivity arcs between the generated and the estimated model; in particular, we analyzed the case of an arc in excess and of an arc missing in the estimated model with respect to the generated one;

3. the same number of connectivity arcs between generated and estimated models, but with an ambiguousness on its orientation.

1.2.3 Evaluation of Performances

In order to evaluate the quality of the estimation performed, the following indexes were computed:

1. the Frobenius norm of the matrix reporting the difference between the values of the estimated (via SEM) and the imposed connections (Relative Error):

$$E_{\text{relative}} = \frac{\sqrt{\sum_{i=1}^{m}\sum_{j=1}^{m}(a_{\ddot{y}} - \hat{a}_{\ddot{y}})^2}}{\sqrt{\sum_{i=1}^{m}\sum_{j=1}^{m}(a_{\ddot{y}})^2}} \qquad (1.13)$$

2. the absolute value of the difference between the estimated parameter and the imposed value on a single particular arc (Single Arc Error):

$$E_{\text{single}} = |a_{\ddot{y}} - \hat{a}_{\ddot{y}}| \qquad (1.14)$$

In order to increase the robustness of the successive statistical analysis, simulations were performed by repeating for 50 runs each connectivity estimation obtained by SEM.

1.2.4 Statistical Analysis

The results obtained were subjected to separate analysis of variance (ANOVA). The main factors of the ANOVA were SNR (with five levels: 1, 3, 5, 10, 100) and LENGTH (with four levels: 60, 190, 310, 610 s). Separate ANOVAs were performed on the error indexes adopted (Relative Error, Single Arc Error). In all the evaluated ANOVAs, the Greenhouse–Gasser corrections for the violation of the spherical hypothesis were used. The post hoc analysis with Duncan's test at $p = 0.05$ statistically significant level was then performed.

1.3   RESULTS OF THE SIMULATIONS

The simulated signals obtained for the different levels of the two SNR and LENGTH factors were analyzed by means of the software LISREL, performing an estimation of the strengths of the connections. Figure 1.1A shows the connection model used in the signal generation and parameter estimation. The arrows represent the existence of a connection directed from the signal z_i toward the signal z_j, while the values on the arcs a_{ij} represent the connection parameters described in Eq. (1.12). The results obtained from 50 repetition of the procedure were subjected to statistical analysis (ANOVA), as reported in detail in the following paragraphs.

1.3.1   Correct Formulation of the Connectivity Model

Figure 1.1 shows the first situation analyzed. A set of signals was generated as described before in order to fit the connectivity pattern shown in Fig. 1.1A. Parameters were estimated on the model shown in Fig. 1.1B, which has exactly the same structure as shown in Fig. 1.1A. We thus test the goodness-of-parameter estimation of model parameters via SEM when no errors are made in the model assumption phase. The appropriate index for this analysis is the Relative Error, as defined in Section 1.2.3, Eq. (1.13). This was computed for each of the 50 runs of the generation–estimation proceedure performed for each level of SNR and LENGTH factors and then subjected to ANOVA, by which was obtained a rather strong statistical significance of both factors for the performance of SEM. In fact, SNR and LENGTH were both highly significant (LENGTH, $F = 288.60$, $p < 0.0001$, SNR, $F = 22.70$, $p < 0.0001$). Figure 1.4A shows the plot of means of Relative Error with respect to the signal length levels, which reveals a decrease of the connectivity estimation error with the increase of the length of the available data. Figure 1.4B shows the plot of means with respect to different SNR levels employed in the simulation. Since the main factors were found to be highly statistically significant, Duncan's post hoc tests at 5% level of significance were then applied. Such tests showed statistically significant differences between all levels of the LENGTH factor, but no statistically significant differences between levels 3, 5, and 10 of the SNR factor.

1.3.2   Hypothesis of a Model with an Arc in Excess or a Missing Arc

Since a perfect formulation of the connectivity model is not always a realistic option, we analyzed several situations in which the connections between the four cortical areas were not correctly assumed in the estimated model.

1.3.3   Arc in Excess

The first situation is described in Fig. 1.2. The SEM parameter estimation was performed on the model shown in Fig. 1.2B, containing an arc which is absent in the imposed pattern (Fig. 1.2A). The aim was to test if the SEM procedure can reject the error made in the model assumption. The appropriate index for this analysis is the Single Arc Error (1.14) on the arc a_{42}, i.e., the one which is not present in the correct model. ANOVA performed on the simulation results showed that both the main factors LENGTH and SNR, have a statistical influence on the ability of SEM to reveal the modeling error. Figure 1.2C and D shows the plot of means with respect to the different levels

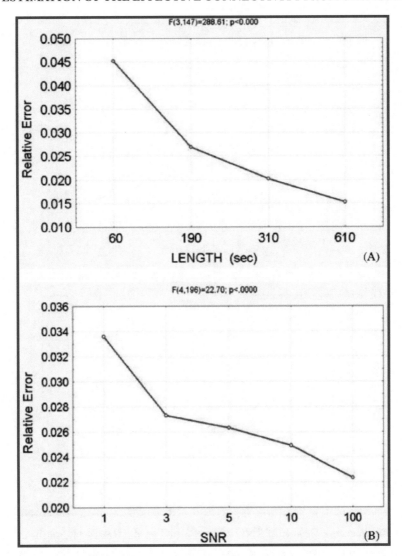

FIGURE 1.4: Correct model. (A) Results of ANOVA performed on the Relative Error. Plot of means with respect to signal LENGTH as a function of time (in seconds). ANOVA shows a high statistical significance for the LENGTH factor ($F = 288.60$, $p < 0.0001$). Duncan's post hoc test performed at 5% level of significance shows statistically significant differences between all levels. (B) Results of ANOVA performed on the Relative Error. Plot of means with respect to SNR. Here too a high statistical influence of the SNR factor on the error in the estimation is seen ($F = 22.70$, $p < 0.001$). Duncan's post hoc test performed at 5% level of significance shows no statistically significant differences between levels 3, 5, and, 10 of the SNR factor. (Published with permission from [6].)

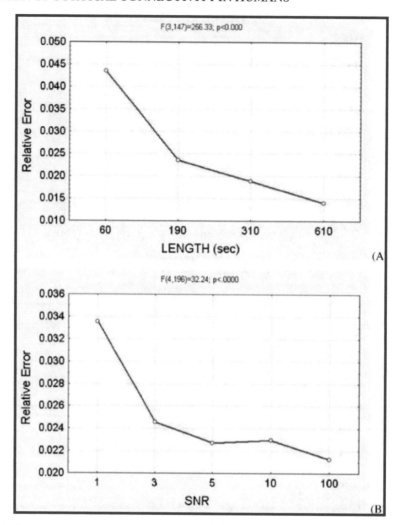

FIGURE 1.5: Arc in excess. Results of ANOVA performed on the Relative Error for the same situation of Fig. 1.3. (A) Plot of means with respect to signal LENGTH as a function of time (in seconds). ANOVA shows a high statistical significance for the LENGTH factor ($F = 256.33$, $p < 0.0001$). Duncan's post hoc test performed at 5% level of significance shows statistically significant differences between all levels. (B) Results of ANOVA performed on the Relative Error. Plot of means with respect to Signal to Noise Ratio. Here too a high statistical influence of the SNR factor on the error in the estimation is seen ($F = 32.24$, $p < 0.001$). Duncan's post hoc test performed at 5% level of significance shows no statistically significant differences between levels 5 and 10 of the SNR factor. (Published with permission from [6].)

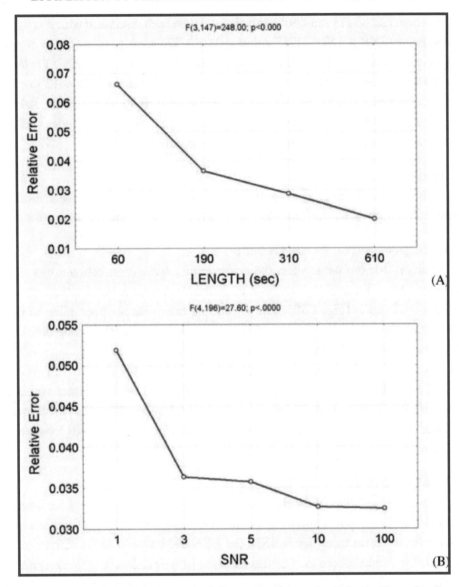

FIGURE 1.6: Arc direction. Results of ANOVA performed on the Relative Error for the same situation as in Fig. 1.5. (A) Plot of means with respect to signal LENGTH as a function of time (in seconds). ANOVA shows a high statistical significance for the LENGTH factor ($F = 248.00, p < 0.0001$). Duncan's post hoc test performed at 5% level of significance shows statistically significant differences between all levels. (B) Results of ANOVA performed on the Relative Error. Plot of means with respect to Signal-to-Noise Ratio. Here too a high statistical influence of the SNR factor on the error in the estimation is seen ($F = 27.60, p < 0.001$). Duncan's post hoc test performed at 5% level of significance shows no statistically significant differences between levels 3, 5, and 10 of the SNR factor. (Published with permission from [6].)

of the main factors LENGTH and SNR. As before, they are both significant with $p < 0.001$ as well as their interaction (SNR × LENGTH) with $p < 0.0001$. Duncan's post hoc test performed at 5% level of significance shows no statistically significant difference between the LENGTH levels of 190 or 310 s as well as between levels 3, 5, 10, and 100 of the main SNR factor. In order to evaluate the influence of the exceeding arc in the model on the global parameter estimation, the Relative Error (1.13) was also computed. Figure 1.5A and B shows the plot of mean of this index with respect to the two main factors, with a level of statistical significance lower than 0.001.

1.3.4   Missing Arc

In this case, the χ^2 statistics test indicates no statistical significance of the estimated model. Hence, the corresponding error values were not computed and no statistical analysis was performed.

1.3.5   Ambiguousness on an Arc Direction

This is a situation that can occur when the existence of a connection between two structures is well known, and there is a need to investigate its direction. Parameters were estimated on a model representing this situation (Fig. 1.3B). The signals had been generated according to the pattern of Fig. 1.3A and the Single Arc Error made on the arc representing the wrong direction (a_{24} in this example) was considered. The statistical analysis performed on the simulation results with the ANOVA reported no statistical significance of the main SNR factor, while the LENGTH factor (EEGTRIAL) is still statistically significant (with $p < 0.0001$). The plot of means in the function of the levels of LENGTH is shown in Fig. 1.3C. Figure 1.6A and B shows the plot of means of the Relative Error with respect to the signal LENGTH levels and to different SNR levels employed in the simulation.

1.4   DISCUSSION

The experimental design adopted for the simulation study aimed at analyzing the most common situations in which the proposed application of SEM technique to EEG data may take place. The levels chosen for the main factor levels SNR and LENGTH as well as the simple errors in the model formulation that have been examined cover the most typical situations that may occur in such analysis. The results obtained have shown a significant statistical influence of the factors considered on SEM performances.

On the basis of the simulations performed, we are now able to answer the questions raised in Section 1.1:

1. There is a statistical influence of a variable SNR level imposed on the high-resolution EEG data on the accuracy of the connectivity pattern estimation. In particular, an SNR $= 3$ seems to be satisfactory in order to obtain good accuracy as there are no significant differences in the performance for higher values.

2. The minimum amount of EEG data necessary to get a usable accuracy of the estimation of connectivity between cortical areas is 190 s of registration (equivalent, for instance, to 380 trials of 500 ms each). However, in this case, an increase in the length of the available EEG data is always related to a decrease in the connectivity estimation error.

3. Different situations, in which the connections between the four cortical areas were not correctly assumed in the estimated model were evaluated in order to analyze their influence on SEM performances. In the first situation, there was a deliberate error in the hypothesized model consisting of the presence of an arc not corresponding to an actual influence between areas. The aim was to test if the SEM procedure can reject the error made in the model assumption and to evaluate the influence of the introduction of such modeling error on the goodness-of-parameter estimation. The analysis of the Single Arc Error on the arc in excess, revealed that an SNR = 3 and an EEG data of 190 s of registration seems to be satisfactory in order to obtain good accuracy. The effect on the global performance of parameter estimation can be inferred by comparing the Relative Error obtained in this situation to the correct one. From Fig. 1.4A and B, compared to Fig. 1.5A and B, it can be seen that the error values remain at the same level in both cases, and the general performance is not decreased by this kind of error. In the second situation analyzed, the voluntary error in the hypothesized model consists of a lack of an arc corresponding to an influence between areas. The performed analysis has not reported any statistical significance, as indicated by the χ^2 to degrees-of-freedom ratio: $\chi^2/df > 1$. This suggests that, in the case of results of this kind, an arc can be added to the putative model in order to decrease the χ^2 to degrees-of-freedom ratio. In the third situation analyzed, the estimated model contained arcs in both directions between two areas, corresponding to a single arc in the model imposed in the signal generation. The Single Arc Error computed on the "wrong direction" arc shows that the error is smaller (less than 1.5% for all factors and levels considered) than in the case of an arc in excess in a single direction in the first situation analyzed. However, it is worth noting that the general performance, as indicated by the Relative Error (Fig. 1.6A and B), is significantly worse in this case than in the case of correct modeling, especially for low values of the LENGTH factor (cf. Fig. 1.4A and B). This means that a simple error such as attributing both directions to a couple of channels causes a significant increase of the error made in the parameter estimation.

In conclusion, the ANOVA results (integrated with the Duncan's post hoc test performed at 5% level of significance indicated a clear influence of different levels of the main factors SNR and LENGTH on the efficacy of the estimation of cortical connectivity via SEM. In particular, it has been noted that at least an SNR = 3 and a LENGTH of the measured cortical data of 190 s are necessary to decrease significantly the errors related to the indexes of the quality adopted.

The simulation study has shown that the ability of SEM to perform a good estimate of connectivity pattern, when connections between the four cortical areas are not correctly assumed, depends on the kind of error made in the model formulation. It seems that the error consisting of a lack of a connection arc is the worst with respect to the parameter estimate, though it can be easily detected by a χ^2 statistical test. Putting in the model an arc not corresponding to an actual influence between areas, on the contrary, does not particularly influence the goodness-of-parameter estimate, and the exceeding arc is attributed a value near to zero. Putting arcs in both directions between two areas, when the influence is directed only from one to the other, causes larger errors in the parameter estimation, though it allows discrimination of the right direction with a precision which does not depend on the signal SNR and which is very high for most levels of the signal LENGTH.

Although the performance seems to be rather good for a correct assumption of the hypothesized model, it decreases when even a simple error is made, depending on the error type. This degradation of the performance seems to indicate the opportunity to use connectivity models that are not too detailed in terms of cortical areas involved as a first step of the network modeling. By using a coarse model of the cortical network to be fitted on the EEG data, there is an increase of the statistical power and a decrease of the possibility of generating an error in a single arc link [4]. In the human study presented in Chapter 2, such an observation was taken into account by selecting a coarse model for the brain areas subserving the task being analyzed. This simplified model does not take complete account of all the possible regions engaged in the task and all the possible connections between them. Elaborate models permitting also cyclical connections between regions can become computationally unstable [5].

In conclusion, the results of the simulation study on SEM performed during this PhD course indicate that the information that a quite accurate estimation of the cortical connectivity patterns can be achieved by using the SEM technique.

REFERENCES

[1] K. Jöreskog and D. Sörbom, LISREL 8.53, software, December 2002. Scientific Software International, Inc. Available: http://www.ssicentral.com, doi:10.1002/hbm.460020107.

[2] B. Horwitz, "The elusive concept of brain connectivity," *Neuroimage*, vol. 19, pp. 466–470, 2003, doi:10.1016/S1053-8119(03)00112-5.

[3] A. R. McIntosh and F. Gonzalez-Lima, "Structural equation modeling and its application to network analysis in functional brain imaging," *Hum. Brain Mapp.*, vol. 2, pp. 2–22, 1994, doi:10.1002/hbm.460020104.

[4] L. Astolfi, et al., "Estimation of the cortical connectivity by high resolution EEG and structural equation modeling: simulations and application to finger tapping data," *IEEE Trans. Biomed. Eng.*, vol. 52, no. 5, pp. 757–768, May 2005, doi:10.1109/TBME.2005.845371.

CHAPTER 2

Estimation of the Functional Connectivity from Stationary Data by Multivariate Autoregressive Methods

2.1 MULTIVARIATE AUTOREGRESSIVE PROCESS

Let Y be a set of cortical waveforms, obtained from several cortical regions of interest (ROI) as described in detail as follows:

$$\mathbf{Y} = [y_1(t), y_2(t), \ldots, y_N(t)]^T \qquad (2.1)$$

where t refers to time and N is the number of cortical areas considered.

Suppose that the following multivariate autoregressive (MVAR) process is an adequate description of the data set Y:

$$\sum_{k=0}^{p} \mathbf{\Lambda}(k)\mathbf{Y}(t-k) = \mathbf{E}(t) \ \text{ with } \ \mathbf{\Lambda}(0) = \mathbf{I} \qquad (2.2)$$

where

- $\mathbf{Y}(t)$ is the data vector in time;
- $\mathbf{E}(t) = [e_1(t), \ldots, e_N]^T$ is a vector of multivariate zero-mean uncorrelated white noise processes;
- $\mathbf{\Lambda}(1), \mathbf{\Lambda}(2), \ldots, \mathbf{\Lambda}(p)$ are the $N \times N$ matrices of model coefficients;
- p is the model order.

In the present study, p was chosen by means of the Akaike information criteria (AIC) for MVAR processes [1] and was used for MVAR model fitting to simulations as well as to experimental signals. It has been noted that, although the sensitivity of MVAR performance depends on the model order, small model order changes do not influence results [2,3].

A modified procedure for the fitting of MVAR on multiple trials was adopted [3–5]. When many realizations of the same stochastic process are available, as in the case of several trials of an event-related potential (ERP) recording, the information from all the trials can be used to increase the reliability and statistical significance of the model parameters. In the present book, the data were in the form of several trials of the same length, both in the simulation and application to real data, as described in detail in the following sections.

After an MVAR model is adequately estimated it becomes the basis for subsequent spectral analysis. To investigate the spectral properties of the examined process, Eq. (2.2) is transformed to the frequency domain

$$\mathbf{\Lambda}(f)\mathbf{Y}(f) = \mathbf{E}(f) \tag{2.3}$$

where

$$\mathbf{\Lambda}(f) = \sum_{k=0}^{p} \mathbf{\Lambda}(k)e^{-j2\pi f \Delta t\, k} \tag{2.4}$$

and Δt is the temporal interval between two samples.

Eq. (2.3) can be rewritten as

$$\mathbf{Y}(f) = \mathbf{\Lambda}^{-1}(f)\mathbf{E}(f) = \mathbf{H}(f)\mathbf{E}(f) \tag{2.5}$$

$\mathbf{H}(f)$ is the transfer matrix of the system, whose element H_{ij} represents the connection between the jth input and the ith output of the system.

2.2   DIRECTED TRANSFER FUNCTION

The directed transfer function (DTF) representing the causal influence of the cortical waveform (estimated in the jth ROI on that estimated in the ith ROI) is defined [6] in terms of elements of the transfer matrix H given by

$$\theta_{ij}^2(f) = |H_{ij}(f)|^2 \tag{2.6}$$

In order to compare the results obtained for cortical waveforms with different power spectra, a normalization can be performed by dividing each estimated DTF by the squared sums of all elements of the relevant row, thus obtaining the so-called normalized DTF [6]

$$\gamma_{ij}^2(f) = \frac{|H_{ij}(f)|^2}{\sum_{m=1}^{N}|H_{im}(f)|^2} \tag{2.7}$$

$\gamma_{ij}(f)$ expresses the ratio of influence of the cortical waveform estimated in the jth ROI on the cortical waveform estimated in the ith ROI with respect to the influence of all the estimated cortical waveforms. Normalized DTF values are in the interval [0, 1], and the normalization condition

$$\sum_{n=1}^{N}\gamma_{in}^2(f) = 1 \tag{2.8}$$

is applied.

From the transfer matrix, we can calculate power spectra $\mathbf{S}(f)$. If we denote by V the variance matrix of the noise $\mathbf{E}(f)$, the power spectrum is defined by

$$S(f) = H(f)\mathbf{V}\mathbf{H}^*(f) \tag{2.9}$$

where the superscript * denotes transposition and complex conjugate.

From $\mathbf{S}(f)$, ordinary coherence can be computed as

$$k_{ij}(f) = \frac{|S_{ij}(f)|^2}{S_{ii}(f)S_{jj}(f)} \tag{2.10}$$

Coherence measures express the degree of synchrony (simultaneous activation) between areas i and j.

2.3   PARTIAL DIRECTED COHERENCE

Partial coherence is another estimator of the relationship between a pair of signals, describing the interaction between areas i and j when the influence due to all $N-2$ time series is discounted. It is defined by the formula

$$|\chi_{ij}(f)|^2 = \frac{|M_{ij}(f)|^2}{M_{ii}(f)M_{jj}(f)} \tag{2.11}$$

where $M_{ij}(f)$ is the minor matrix obtained by removing ith row and jth column from the spectral matrix S.

In 2001, Baccalà proposed the following factorization:

$$\chi_{ij}(f) = \frac{\Lambda_i^*(f)V^{-1}\Lambda_j(f)}{\sqrt{\left(\Lambda_i^*(f)V^{-1}\Lambda_i(f)\right)\left(\Lambda_j^*(f)V^{-1}\Lambda_j(f)\right)}} \tag{2.12}$$

where $\Lambda_n(f)$ is the nth column of the matrix $\mathbf{\Lambda}(f)$. This led to the definition of partial directed coherence (PDC), [7]

$$\pi_{ij}(f) = \frac{\Lambda_{ij}(f)}{\sqrt{\sum_{k=1}^{N}\Lambda_{ki}(f)\Lambda_{kj}^*(f)}} \tag{2.13}$$

The PDC from j to i, $\pi_{ij}(f)$, describes the directional flow of information from the activity in the ROI $s_j(n)$ to the activity in $s_i(n)$, whereupon common effects produced by other ROIs $s_k(n)$ on the latter are subtracted, leaving only a description that is specifically from $s_j(n)$ to $s_i(n)$.

The PDC values are in the interval $[0, 1]$, and the normalization condition

$$\sum_{n=1}^{N}|\pi_{ni}(f)|^2 = 1 \tag{2.14}$$

is verified. According to this condition, $\pi_{ij}(f)$ represents the fraction of the time evolution of ROI j directed to ROI i, compared to all of j's interactions with other ROIs.

For both DTF and PDC high values in a frequency band represent the existence of an influence between any given pair of areas in the data set. However, an important difference is that PDC does not involve the inversion of matrix $\mathbf{\Lambda}$. This leads to several points. In fact, an analysis of the definition of DTF reveals that, due to this matrix inversion, it is a linear combination of both the direct influence from one area to the other and the influence mediated by other areas along various cascade pathways [8]. This becomes immediately clear from an example. Given a three-region model, the nonnormalized DTF from area 1 to area 2 is given by

$$\theta_{21}^2(f) = |H_{21}(f)|^2 = \frac{|\Lambda_{21}(f)\Lambda_{33}(f) - \Lambda_{31}(f)\Lambda_{23}(f)|}{|\mathbf{\Lambda}(f)|^2} \qquad (2.15)$$

From this formula it can be noted that even if the direct influence from area 1 to area 2, $\Lambda_{21}(f)$, is zero, $\theta_{21}^2(f)$ may still be different from zero since there is an influence from 1 to 3 ($\Lambda_{31}(f)$) and from 3 to 2 ($\Lambda_{23}(f)$). The link between 1 and 2 will be indicated by DTF as a causal pathway if all the causal influences along the way are nonzero.

The PDC, due to the lack of the matrix inversion, behaves differently. It indicates only the existence of a direct causal influence from area 1 to area 2. If no direct influence exists, PDC_{21} is virtually zero.

2.4   DIRECT DTF

In order to distinguish between direct and cascade flows in DTF, the direct DTF (dDTF) was introduced [9]. It is defined by multiplying the full frequency directed transfer function (ffDTF) given by

$$\eta_{ij}^2(f) = \frac{\left|H_{ij}(f)\right|^2}{\sum_f \sum_{m=1}^{k} |H_{im}(f)|^2} \qquad (2.16)$$

by the partial coherence defined in Eq. (2.11). The dDTF from area j to area i is defined as

$$\delta_{ij}(f) = \chi_{ij}(f)\eta_{ij}(f) \qquad (2.17)$$

This function describes only the direct relations between channels. The denominator of the ffDTF function (2.16) does not depend on the frequency.

2.5   SIMULATION STUDY

The experimental design involved the following steps:

- *Generation of a set of test signals simulating cortical average activations*: Several sets of simulated data were generated in order to fit a predefined connectivity model and to respect imposed levels of the signal-to-noise ratio (factor SNR) and the length of the data (factor LENGTH).

The data were in the form of multiple trials, and the factor LENGTH indicates the total length of all trials.

- *Estimation of the cortical connectivity pattern obtained under different conditions*: The estimators used were DTF, PDC, and dDTF.

- *Computation of indices of connectivity estimation performance*: These indices were error functions describing the error in the connectivity estimation for the whole pattern and for each single arc. A comparison between the value estimated for the direct and indirect arcs was also performed.

- *Statistical analysis*: Analysis of variance (ANOVA) of the results of the simulations was performed to study the effects of the factors SNR and LENGTH on the recovery of the connectivity pattern resulting from the different methods.

2.5.1   Signal Generation

The connectivity model used in the generation of test signals is shown in Fig. 2.1. It involves five areas linked by both direct and indirect pathways. For example, ROIs 1 and 2 are linked by a *direct* path directed from 1 to 2. ROIs 1 and 5 are not connected by any direct arc but are linked by an *indirect* path from 1 to 2 and from 2 to 5. ROIs 4 and 5 are not linked by either a direct or an indirect arc. As described, these situations are rather different with respect to the estimates obtained by multivariate methods based on MVAR models. In particular, the model shown in Fig. 2.1 has 7 direct arcs, 2 indirect arcs, and 11 "null" arcs (i.e., 11 pairs of ROIs are not linked, either directly or indirectly).

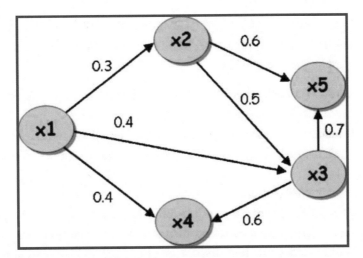

FIGURE 2.1: Connectivity model imposed in the generation of simulated signals. Values on the arrows represent the connections strengths. (Published with permission from [10].)

The simulated signals were obtained starting from a neural mass model of an ROI, fitted to a real cortical estimation of the average activity in an ROI. Signal x_1 was a waveform generated by a model of three neural populations arranged in parallel. Each population simulates neural activity in a specific frequency band: 4–12 Hz, 12–30 Hz, and 30–50 Hz. The model of each population is based on equations proposed by Wendling et al. [11]. The basic idea behind this model is that the oscillations derive from the interactions of pyramidal neurons with three other local neural subsets: excitatory interneurons, slow inhibitory interneurons, and fast inhibitory interneurons. Parameters of the three populations (time constants and synaptic gains) have been set by using an automatic best-fitting procedure to mimic the entire power spectrum density of cortical activity in an ROI.

Subsequent signals $x_2(t)$ to $x_5(t)$ were iteratively obtained according to the imposed connectivity scheme (Fig. 2.1), by adding to signal x_j contributions from the other signals, delayed by intervals τ_{ij}, and amplified by factors a_{ij} and uncorrelated Gaussian white noise. Coefficients of the connection strengths were chosen in a range of realistic values as observed in studies that applied other connectivity estimation techniques, such as structural equation modeling to several memory, motor, and sensory tasks [12,13]. The values used for the connection strengths are given in the legend for Fig. 2.1. The data were generated using different delay schemes on the connectivity pattern imposed. The values used for the delay from the ith ROI to the jth (τ_{ij}) ranged from one sample up to $p - 2$, where p is the order of the MVAR model used. These schemes were chosen in order to cover a variety of situations to represent the effect of different delay conditions.

Generation of simulated data was repeated under the following combinations of conditions:

SNR factor levels $= [1, 3, 5, 10]$;

LENGTH factor levels $= [2,500, 6,750, 11,250, 15,000, 20,000]$ data samples, corresponding to a signal length (in seconds) of $[10, 27, 45, 60, 80]$, in the form of several trials of the same length, at a sampling rate of 250 Hz.

The levels chosen for both SNR and LENGTH factors cover a typical range for cortical activity estimated from ERPs with high resolution EEG techniques.

The MVAR model was estimated by means of the Nuttall–Strand method, or multivariate Burg algorithm, which has been demonstrated to provide the most accurate results [14–16].

2.5.2   Evaluation of Performance

A statistical evaluation of the performance of the different estimators required a precise definition of an error function, describing the goodness of the pattern recognition. This was achieved by focusing on the MVAR model structure described in Eq. (2.1) and comparing it with the signal generation scheme. The elements of matrices $\Lambda(k)$ of MVAR model coefficients can be related to the coefficients used in the signal generation and are different from zero only for $k = \tau_{ij}$, where τ_{ij}

is the delay chosen for each pair ij of ROIs and for each direction among them. In particular, for the independent reference source waveform $x_1(t)$, an autoregressive model of the same order of the MVAR was estimated, with coefficients $a_{11}(1), \ldots, a_{11}(p)$ corresponding to the elements $\Lambda_{11}(1), \ldots, \Lambda_{11}(p)$ of the MVAR coefficients matrix. Thus, with the estimation of the MVAR model parameters, we aimed to recover the original coefficients $a_{ij}(k)$ used in signal generation. In this way, reference functions were computed for each of the estimators on the basis of the signal generation parameters. The error function was then computed as the difference between these reference functions and the estimated ones (both averaged in the frequency band of interest).

To evaluate the performance in retrieving connections between areas, we used the Frobenius norm of the matrix reporting the differences between the values of the estimated and the imposed connections (Relative Error)

$$E_{\text{relative}} = \frac{\sqrt{\sum_{i=1}^{m}\sum_{j=1}^{m}\left(\bar{\zeta}_{ij}(f_1,f_2) - \bar{\tilde{\zeta}}_{ij}(f_1,f_2)\right)^2}}{\sqrt{\sum_{i=1}^{m}\sum_{j=1}^{m}\left(\bar{\zeta}_{ij}(f_1,f_2)\right)^2}} \qquad (2.18)$$

where $\bar{\tilde{\zeta}}_{ij}(f_1,f_2)$ is the mean value of the estimator function estimated in the frequency band (f_1,f_2), and $\bar{\zeta}_{ij}(f_1,f_2)$ is the mean value of the reference functions obtained from the generation model in the same frequency band. Here, $\bar{\zeta}_{ij}$ can be either DTF or PDC.

Simulations were performed by repeating each generation-estimation procedure 50 times in order to increase the robustness of the successive statistical analysis.

2.5.3   Statistical Analysis

The results obtained were subjected to separate ANOVA study. The first analysis was a three-way ANOVA examining the effect of SNR, LENGTH, and the different methods used to estimate the cortical connectivity (METHODS) on the error for the entire connectivity model estimated. The "within" main factors of the ANOVAs were as follows:

SNR (with four levels: 1, 3, 5, 10),

LENGTH (with six levels: [2,500, 6,750, 11,250, 15,000, 20,000] data samples, corresponding to a signal length (in seconds) of [10, 27, 45, 60, 80], in three trials of the same length, at a sampling rate of 250 Hz) and METHODS (with two levels: DTF and PDC).

The dependent variable was the Relative Error defined in (2.18). The Greenhouse–Geisser correction for the violation of the spherical hypothesis was used. Duncan's post hoc analyses at $p = 0.05$ significance level were then performed.

As explained above, the presence of indirect paths in the network (i.e., a path linking a node to another node on the network not directly but through one or more intermediate nodes) is a critical situation for MVAR-based estimators of causality relations. For this reason, particular attention was paid to the analysis of the estimation error in such indirect relationships. Another ANOVA was performed where the values estimated on the indirect arcs by the three methods, DTF, PDC, and dDTF were compared to the average value of the parameters estimated on the arcs actually present in the generation model. The dependent variable was the absolute level of the connectivity estimates, while the independent factors were METHODS, LENGTH, SNR, and PATHS. The first three main factors had the same levels used before in the other ANOVAs, and the main factor PATHS had three levels, describing the value of the dependent variable for the "indirect" links moving from the area x_1 to the area x_5, that from the area x_2 to the area x_4, and the average value estimated on nonzero arcs in the model. The Greenhouse–Geisser correction was used also in this case.

2.6   RESULTS

Several sets of signals were generated as described in the previous section in order to fit a predefined connectivity pattern involving five cortical areas (shown in Fig. 2.1). The graph depicts the flow of information from area x_1 toward areas x_2–x_5. This connectivity model contains two indirect paths, by which the signal is transmitted to a destination only by indirect relationships with no direct link between the source area and the target one (from area x_2 to area x_4 through x_3, and from area x_1 to area x_5 through x_2 and x_3, by several different paths: $x_1 \rightarrow x_3 \rightarrow x_5$ and $x_1 \rightarrow x_2 \rightarrow x_3 \rightarrow x_5$).

A multivariate autoregressive model of order 10 was fitted to each set of simulated data, which were in the form of about 30 trials of the same length. The procedure of signal generation and connectivity estimation for the different methods was carried out 50 times for each level of the factors SNR and LENGTH for increasing the robustness of the subsequent statistical analysis. The index of performance, i.e., the Relative Error [Eq. (2.18)], and the estimated value on direct/indirect arcs were computed for each generation-estimation procedure and then subjected to two ANOVAs.

In the first ANOVA, the dependent variable was the Relative Error, representing the average error for the entire connectivity pattern estimated. Results revealed a strong influence of the main factors SNR ($F = 205, p < 0.0001$), METHODS ($F = 1190, p < 0.0001$), and LENGTH ($F = 1644, p < 0.0001$), as well as the SNR × METHOD interaction ($F = 56, p < 0.0001$), on the Relative Error. Figure 2.2 shows the influence of the levels of the main factors LENGTH and SNR on the Relative Error in the entire connectivity graph for each estimator. The errors resulting from DTF, evaluated with respect to the theoretical values that represent the ideal information obtainable by this indicator, were smaller than those resulting from PDC for every level of SNR and LENGTH. This indicates that DTF is more robust with respect to noise and the amount of data available for its estimation. In particular, Fig. 2.2 shows that for each increase in the length of the recordings, there is a constant decrease in estimation error. The influence of the factor SNR is weaker. Post hoc tests revealed that there were no significant differences between levels 3, 5, and 10 of the SNR factor. The bar on each point represents the 95% confidence interval of the mean errors computed across the simulations.

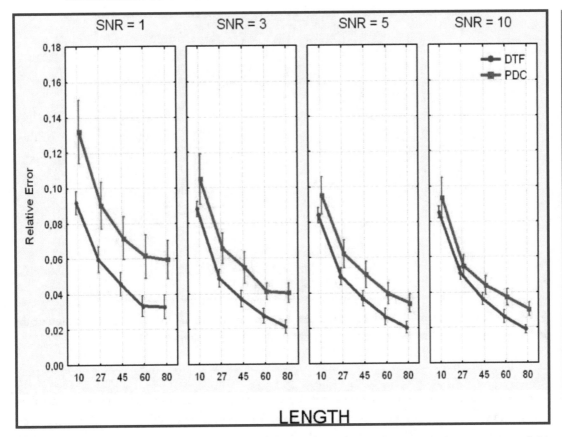

FIGURE 2.2: Results of the ANOVA performed on the relative error made in the estimation of the connectivity flows. The diagram shows the influence of the different levels of the main factors SNR and LENGTH on the estimation of the correct flows in the connection graph employed for the simulation, for the two estimators DTF and PDC. The bar on each point represents the 95% confidence interval of the mean errors computed across the simulations. Duncan's post hoc test (performed at 5%) showed no significant difference between levels 3, 5, and 10 of factor SNR. (Published with permission from [10].)

Particular attention was paid to the ability of the different estimators to distinguish between direct and indirect causality flows. The values estimated for the two indirect pathways in the model were compared to the average value obtained for the direct arcs present in the connectivity model imposed. The values of the connections imposed between cortical areas ranged from 0.3 to 0.7. The two indirect paths analyzed were arc $1 \rightarrow 5$ and arc $2 \rightarrow 4$, indicated by big arrows in Fig. 2.3. As expected, DTF was not always able to attribute a zero value to a nondirect arc. In contrast, PDC and dDTF correctly recognized the indirect path in all cases, estimating values close to zero in these instances (see Fig. 2.4). However, the results obtained by dDTF for the direct arcs were also very

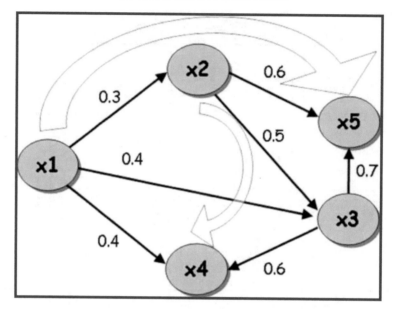

FIGURE 2.3: Connectivity model imposed on the simulated signals. The thick arrows represent the indirect pathways linking the cortical areas. (Published with permission from [10].)

small when compared to the other methods. Post hoc tests revealed no differences between values obtained by dDTF on the average of direct arcs and the values obtained by the same method on indirect arcs. The ANOVA revealed a strong influence of the main factors SNR ($F = 20, p < 0,0001$), METHOD ($F = 2980000, p < 0,0001$), and LENGTH ($F = 493, p < 0,0001$). Furthermore, all possible interactions of the main factors were significant, with F values not below 3 and p always below 0.001.

2.7   DISCUSSION

The present study examines the performance of three different techniques commonly used to assess information flows between scalp electrodes and local field potentials [7–9] on simulated and real cortical waveforms obtained via the linear inverse problem solution, using the realistic head volume conductor models and high-density EEG recordings. The spatial resolution provided by the techniques presented here has been previously characterized in a series of simulation studies using the present ROI analysis approach [17–20]. The simulation studies involved realistic head models, high-density EEG and MEG setups (with 64 and 128 electrodes as well as 143 magnetic sensors), and standard levels of SNRs (1, 3, 5, 10, 100). The returned errors are lower than 5% for the estimation of the shape (via the cross-correlation measure) and energy (via the Relative Error measurement) of the simulated cortical waveforms. These figures assure that the estimation

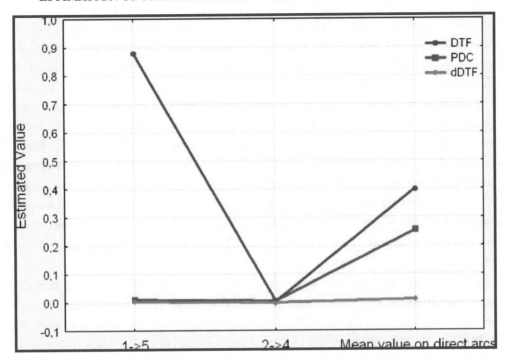

FIGURE 2.4: Average connectivity values estimated on two indirect links (1 → 5 and 2 → 4) and for all the other existing arcs for the networks by the three methods DTF, PDC, and dDTF during all the simulations. (Published with permission from [10].)

of the cortical current density using high spatial sampling and realistic head models is a reliable process that allows a precise reconstruction of the cortical waveforms in a variety of experimental situations.

All the techniques investigated in the present study are based on the Granger theory and MVAR models. The approach using the DTF, PDC, and dDTF techniques has the advantage of providing connectivity links that can be interpreted in the sense of Granger causality, which includes a concept of directionality. Other techniques have been presented in the literature for the evaluation of functional connectivity of EEG/MEG data. For instance, the technique called dynamic imaging of coherent sources (DICS) [21,22], which uses a spatial filter and a realistic head model, has been recently introduced and employed to assess connectivity between cortical areas from MEG data [21,22]. This technique has the advantage, when compared to the DTF, PDC, and dDTF methods investigated here, of a direct mathematical characterization of its spatial resolution of the point spread function [22]. However, spectral coherence or DICS techniques do not return directly the direction of the flow between cortical areas, though in the latter case, DICS is usually coupled with another technique able to estimate such directional flow, e.g., directionality index [23]. The main difference between DICS and the approach presented here is in the estimation of cortical activity, specifically the fact that the DICS technique is a beamformer [24]. Differences in the performance of beamformers and weighted minimum norm linear inverse techniques depend on the particular experimental setup used. In particular, it has been shown in simulation studies that EEG/MEG beamformers can reconstruct rather precisely the spatial location and the time series of neuronal sources when they exhibit a transient correlation in time [25]. If, however, the correlation between source activities exceeds 30–40% of the total duration of the period over which the beamformer weights are computed, effects of temporal distortion and signal cancellation will be observed [21,26,27]. This suggests that, if the beamformer is obtained using covariance time windows that are long enough with respect to the duration of transient linear interaction between sources, it will return an accurate estimate of spatiotemporal source activity. However, if the covariance windows are sufficiently long, the portion of the stimulus will be small in comparison with the baseline state, and a decrease in the SNR will occur. This would ultimately result in a loss of spatial resolution [21,26].

An interesting issue is related to the possibility of applying the connectivity estimators to multimodal neuroelectric and hemodynamic data (i.e., from EEG/MEG and fMRI measurements). Recent simulation and experimental studies [3,19,28,29] stated that the multimodal integration of EEG, MEG, and fMRI improves the quality of the cortical estimation when compared to any single modality alone. This was obtained by using MEG and fMRI estimation [28,29], EEG and fMRI [3,19,30], or EEG and MEG [18,31]. It is reasonable to expect that, with the improvement of the quality of the cortical estimation given by multimodal integration, the estimation of connectivity could improve as well. However, the effect of multimodal integration in terms of connectivity is not yet addressed in literature, due to the lack of a precise model of electrovascular coupling. It is also worth noting that all the techniques (EEG, MEG, and fMRI) can detect the activity of a particular set of neural sources and are blind to others. For instance, the activity of stellate neurons in the cortex can be detected by fMRI because of their metabolic demand, but not by neuroelectromagnetic measurements to which they are invisible due to the closed field they generate. In contrast, transient (milliseconds) synchronous activity of a small subset of neurons can be detected by EEG and MEG but is invisible to fMRI [32]. Hence, the use of multimodal integration can provide information about cortical activity that moves beyond that offered by a single technique. This is an important point in favor of multimodal integration of EEG, MEG, and fMRI, especially in the perspective of the estimation of functional cortical connectivity.

We performed a series of simulations to evaluate the use of connectivity estimators on test signals generated to simulate the average electrical activity of cerebral cortical regions, as it can be estimated from high resolution EEG recordings gathered under different conditions of noise and length of the recordings. The information on performances, limits of applicability, and range of errors under different levels of the several factors that are of interest in normal EEG recordings were inferred from statistical analysis (ANOVA and Duncan's post hoc tests on the Relative Error and

Single Arc Error). The values used for the strength coefficients in simulations are consistent with the ones estimated in previous studies for a large sample of subjects performing memory, motor, and sensory tasks [12,13].

The simulations provided the following answers to the questions raised in the Introduction chapter:

1. Decreased SNR impairs the accuracy of the connectivity pattern estimation obtained by the DTF, PDC, and dDTF estimators.

2. The length of the EEG recordings has a reliable effect on the accuracy of connectivity pattern estimations. A length corresponding to 27 s of nonconsecutive recordings with an SNR of at least 3 ensures that connectivity patterns can be accurately recovered with an error below 7% for PDC and 5% for DTF.

3. The error variance observed for the DTF estimator is lower than that for PDC or dDTF. However, DTF has the highest bias in the estimate of the connectivity pattern as it includes values for the indirect paths that were not generated in the simulations. It has been noted that there is a higher bias for dDTF than for PDC, as the first estimator often removes from the estimated connectivity pattern some direct paths that were present in the original modeling. In this respect, PDC is characterized by lower bias in the estimation of connectivity patterns under the present conditions of SNR and LENGTH.

In conclusion, results indicated a clear influence of different levels of SNR and LENGTH on the efficacy of the estimation of cortical connectivity in each of the methods. In particular, it has been noted that an SNR equal to or greater than 3 and an overall LENGTH of the estimated cortical data of 6750 data samples (27 s at 250 Hz), even in several short trials, are sufficient to significantly decrease the errors on the indices of quality adopted in this study. These conditions are common in recordings of event-related activity in humans. These recordings are usually characterized by SNR ranging from 3 (movement-related potentials) to 10 (sensory evoked potentials) [33].

The present simulations allowed an evaluation of the level of error expected for different arcs, related to direct or indirect pathways. In addition, comparisons were made between the relative errors obtained on single arcs characterized by only direct connections and those related to multiple paths between the source and the target. The results showed that the error was generally greater when the signal was transmitted to a destination by more than one path. This result (not presented here) is in agreement with the results of the study on indirect paths, according to which such transmission may induce an error in the MVAR estimation.

ANOVA was also performed on the error values obtained for different delay schemes imposed during the signal generation for covering a variety of situations. The ANOVA results indicated that there was no significant influence of the delay on the performances of the methods.

The information obtained from the simulations was used to evaluate the applicability of these methods to actual event-related recordings. The ERP signals, from a Stroop task, showed an SNR between 3 and 5 in the five subjects examined. Therefore, according to the simulation results, a small amount of error in the estimation of cortical connectivity patterns was expected.

All three estimators provided *directional* information (i.e., each of them allowed the establishment of the direction of information flow between two cortical areas) and *directed* information (i.e., they discriminated between direct and indirect connection paths). This information is not available using other techniques to assess the coupling between signals such as standard coherence (which lacks directionality). An evaluation of several methods for the computation of the functional connectivity between EEG/MEG signals was recently performed [34]. It was concluded that, although nonlinear methods such as mutual information, nonlinear correlation, and generalized synchronization [35–37] might be preferred when studying EEG broadband signals that are sensitive to dynamic coupling and nonlinear interactions expressed over many frequencies, the linear measurements (like those presented here) afford a rapid and straightforward characterization of functional connectivity.

REFERENCES

[1]   H. Akaike, "A new look at statistical model identification", IEEE Trans Automat Control AC-19, 1974, pp. 716–723, doi:10.1109/TAC.1974.1100705.

[2]   P. J. Franaszczuk, K. J. Blinowska, and M. Kowalczyk, "The application of parametric multichannel spectral estimates in the study of electrical brain activity," *Biol. Cybern.*, vol. 51, pp. 239–247, 1985, doi:10.1007/BF00337149.

[3]   F. Babiloni, F. Cincotti, C. Babiloni, F. Carducci, A. Basilisco, P. M. Rossini, D. Mattia, L. Astolfi, L. Ding, Y. Ni, K. Cheng, K. Christine, J. Sweeney, and B. He, "Estimation of the cortical functional connectivity with the multimodal integration of high resolution EEG and fMRI data by Directed Transfer Function," *Neuroimage*, vol. 24, no. 1, pp. 118–131, Jan. 1 2005, doi:10.1016/j.neuroimage.2004.09.036.

[4]   M. Ding, S. L. Bressler, W. Yang, and H. Liang Ding, "Short-window spectral analysis of cortical event-related potentials by adaptive multivariate autoregressive modeling: data preprocessing, model validation, and variability assessment," *Biol. Cybern.*, vol. 83, pp. 35–45, 2000, doi:10.1007/s004229900137.

[5]   L. Astolfi, F. Cincotti, D. Mattia, C. Babiloni, F. Carducci, A. Basilisco, P. M. Rossini, S. Salinari, L. Ding, Y. Ni, B. He, and F. Babiloni, "Assessing cortical functional connectivity by linear inverse estimation and directed transfer function: simulations and application to real data," *Clinical Neurophysiology*, vol. 116 no. 4, pp. 920–932, Apr. 2005b, doi:10.1016/j.clinph.2004.10.012.

[6]   M. Kaminski, K. Blinowska, "A new method of the description of the information flow in the brain structures," *Biol. Cybern.*, vol. 65, pp. 203–210, 1991, doi:10.1007/BF00198091.

[7]   L. A. Baccalà, and K. Sameshima, "Partial Directed Coherence: a new concept in neural structure determination," *Biol. Cybern.*, vol. 84, pp. 463–474, 2001, doi:10.1007/PL00007990.

[8]   M. Kaminski, M. Ding, W. A. Truccolo, and S. Bressler, "Evaluating causal relations in neural systems: Granger causality, directed transfer function and statistical assessment of significance," *Biol. Cybern.*, vol. 85, pp. 145–157, 2001, doi:10.1007/s004220000235.

[9]   A. Korzeniewska, M. Manczak, M. Kaminski, K. Blinowska, and S. Kasicki, "Determination of information flow direction between brain structures by a modified Directed Transfer Function method (dDTF)," *Journal of Neuroscience Methods*, vol. 125, pp. 195–207, 2003, doi:10.1016/S0165-0270(03)00052-9.

[10] L. Astolfi, F. Cincotti, D. Mattia, M. G. Marciani, L. A. Baccala, F. De Vico Fallani, S. Salinari, M. Ursino, M. Zavaglia, L. Ding, J. C. Edgar, G. A. Miller, B. He, and F. Babiloni, "A

comparison of different cortical connectivity estimators for high resolution EEG recordings," *Human Brain Mapping*, vol. 28, no. 2, pp. 143–157, Feb 2007, doi:10.1002/hbm.20263.

[11] F. Wendling, F. Bartolomei, J. J. Bellanger, and P. Chauvel, "Epilepsy fast activity can be explained by a model of impaired GABAergic dendritic inhibition," *Eur. J. Neurosci.*, vol. 15, no. 9, pp. 1499–1508, May 2002, doi:10.1046/j.1460-9568.2002.01985.x.

[12] C. Buchel, and K. J. Friston, "Modulation of connectivity in visual pathways by attention: cortical interactions evaluated with structural equation modeling and fMRI," *Cereb. Cortex*, vol. 7, no. 8, pp. 768–778, 1997, doi:10.1093/cercor/7.8.768.

[13] L. Fa-Hsuan, *Spatio temporal brain imaging and modeling.* PhD Thesis, MIT Press, December 2003.

[14] S. L. Marple, *Digital Spectral Analysis with Applications*, Prentice Hall, 1987.

[15] M. S. Kay, *Modern Spectral Estimation*, Prentice Hall, 1988.

[16] A. Schlögl (2003). Comparison of Multivariate Autoregressive Estimators. Available online at: http://www.dpmi.tugraz.ac.at/~schloegl/publications/TR_MVARcomp201.pdf.

[17] F. Babiloni, C. Babiloni, L. Locche, F. Cincotti, P. M. Rossini, and F. Carducci, "High-resolution electroencephalogram: source estimates of Laplacian-transformed somatosensory-evoked potentials using a realistic subject head model constructed from magnetic resonance images," *Med. Biol. Eng. Comput.* vol. 38, no. 5, pp. 512–519, Sep. 2000, doi:10.1007/BF02345746.

[18] F. Babiloni, F. Carducci, F. Cincotti, C. Del Gratta, V. Pizzella, G. L. Romani, P. M. Rossini, F. Tecchio, and C. Babiloni, "Linear inverse source estimate of combined EEG and MEG data related to voluntary movements," *Human Brain Mapping*, vol. 14, no. 3, 2001, doi:10.1002/hbm.1052.

[19] F. Babiloni, C. Babiloni, F. Carducci, G. L. Romani, P. M. Rossini, L. M. Angelone, and F. Cincotti, "Multimodal integration of high-resolution EEG and functional magnetic resonance imaging data: a simulation study," *Neuroimage*, vol. 19, no. 1, pp. 1–15, May 2003, doi:10.1016/S1053-8119(03)00052-1.

[20] F. Babiloni, C. Babiloni, F. Carducci, G. L. Romani, P. M. Rossini, A. Basilisco, S. Salinari, L. Astolfi, and F. Cincotti, "Solving the neuroimaging puzzle: the multimodal integration of neuroelectromagnetic and functional magnetic resonance recordings," *Suppl. Clin. Neurophysiol.*, vol. 57, pp. 450–457, 2004c.

[21] J. Gross, J. Kujala, M. Hämäläinen, L. Timmermann, A. Schnitzler, R. Salmelin, "Dynamic imaging of coherent sources: studying neural interactions in the human brain," *Proc. Natl. Acad. Sci. USA*, vol. 98, no. 2, pp. 694–699, 2001, doi:10.1073/pnas.98.2.694.

[22] J. Gross, L. Timmermann, J. Kujala, R. Salmelin, and A. Schnitzler, "Properties of MEG tomographic maps obtained with spatial filtering," *NeuroImage* vol. 19, pp. 1329–1336, 2003, doi:10.1016/S1053-8119(03)00101-0.

[23] M. G. Rosenblum, and A. S. Pikovsky, "Detecting direction of coupling in interacting oscillators," *Phys. Rev., E Stat. Nonlinear Soft. Matter. Phys.*, vol. 64, p. 045202, 2001, doi:10.1103/PhysRevE.64.045202.

[24] M. X. Huang, J. Shih, R. R. Lee, D. L. Harrington, R. J. Thoma, M. P. Weisend, F. M. Hanlon, K. M. Paulson, T. Li, K. Martin, G. A. Miller, and J. M. Cañive, "Commonalities and differences among vectorized beamformers in electromagnetic source imaging," *Brain Topography*, vol. 16, pp. 139–158, 2003, doi:10.1023/B:BRAT.0000019183.92439.51.

[25] A. Hadjipapas, A. Hillebrand, I. E. Holliday, K. D. Singh, and G. R. Barnes, "Assessing interactions of linear and nonlinear neuronal sources using MEG beamformers: a proof of concept," *Clin. Neurophysiol.*, vol. 116, no. 6, pp. 1300–1313, Jun 2005. Epub 2005 Mar 28, doi:10.1016/j.clinph.2005.01.014.

[26] VanVeen, B. D. vanDrongelen, W. Yuchtman, and M. Suzuki, "A Localization of brain electrical activity via linearly constrained minimum variance spatial filtering," *IEEE Trans. Biomed. Engl.*, vol. 44, pp. 867–880, 1997, doi:10.1109/10.623056.

[27] K. Sekihara, S. Nagarajan, D. Poeppel, and A. Marantz, "Performance of an MEG adaptive-beamformer technique in the presence of correlated neural activities: effects on signal intensity and time-course estimates," *IEEE Trans. Biomed. Eng.*, vol. 49, pp. 1534–1546, 2002, doi:10.1109/TBME.2002.805485.

[28] A. K. Liu, J. W. Belliveau, and A. M. Dale, "Spatiotemporal imaging of human brain activity using functional MRI constrained magnetoencephalography data: Monte Carlo simulations," *Proc. Nat. Acad. Sc.*, vol. 95, no. 15, pp. 8945–8950, 1998, doi:10.1073/pnas.95.15.8945.

[29] A. Liu Dale, B. Fischl, R. Buckner, J. W. Belliveau, J. Lewine, and E. Halgren, "Dynamic statistical parametric mapping: combining fMRI and MEG for high-resolution imaging of cortical activity," *Neuron*, vol. 26, pp. 55–67, 2000, doi:10.1016/S0896-6273(00)81138-1.

[30] F. Babiloni, D. Mattia, C. Babiloni, L. Astolfi, S. Salinari, A. Basilisco, P. M. Rossini, M. G. Marciani, and F. Cincotti, "Multimodal integration of EEG, MEG and fMRI data for the solution of the neuroimage puzzle," *Magnetic Resonance Imaging*, vol. 22, no. 10, pp. 1471–1476, Dec 2004a, doi:10.1016/j.mri.2004.10.007.

[31] F. Babiloni, C. Babiloni, F. Carducci, G. L. Romani, P. M. Rossini, L. M. Angelone, and F. Cincotti, "Multimodal integration of EEG and MEG data: a simulation study with variable signal-to-noise ratio and number of sensors," *Hum. Brain Mapp.*, vol. 22, no. 1, pp. 52–62, May 2004b, doi:10.1002/hbm.20011.

[32] P. L. Nunez, *Neocortical Dynamics and Human EEG Rhythms*, New York: Oxford University Press, 1995.

[33] D. Regan, *Human Brain Electrophysiology. Evoked Potentials and Evoked Magnetic Fields in Science and Medicine*, New York: Elsevier Press, 1989.

[34] O. David, D. Cosmelli, and K. J. Friston, "Evaluation of different measures of functional connectivity using a neural mass model," *NeuroImage*, vol. 21, pp. 659–673, 2004, doi:10.1016/j.neuroimage.2003.10.006.

[35] M. S. Roulston, "Estimating the errors on measured entropy and mutual information," *Physica D* vol. 125, pp. 285–294, 1999, doi:10.1016/S0167-2789(98)00269-3.

[36] C. J. Stam, and B. W. van Dijk, "Synchronization likelihood: an unbiased measure of generalized synchronization in multivariate data sets," *Physica, D*, vol. 163, pp. 236–251, 2002, doi:10.1016/S0167-2789(01)00386-4.

[37] C. J. Stam, M. Breakspear, A. M. van Cappellen van Walsum, and B. W. van Dijk, "Nonlinear synchronization in EEG and whole head MEG recordings of healthy subjects," *Hum. Brain Mapp.*, vol. 19, no. 2, pp. 63–78, 2003, doi:10.1002/hbm.10106.

CHAPTER 3

Estimation of Cortical Activity by the use of Realistic Head Modeling

3.1 THE PROBLEMS OF CONVENTIONAL ELECTROENCEPHALOGY RECORDINGS

Among the different noninvasive brain imaging techniques, electroencephalography (EEG) and magneto encephalography (MEG) alone directly reflect neuronal firing and exhibit a remarkable temporal resolution (in milliseconds) despite a poor spatial resolution (of the order of fewsquare centimeters). This lack of spatial resolution occurs essentially in the case of EEG because of the spread of brain signals due to the low conductivity of the skull and the rather low signal-to-noise ratio of the data [1]. These potentials originate mainly in the radially oriented cortical pyramidal neurons. The potential distribution arising from these sources is quite wide over the scalp surface because of the different conductivities of cerebrospinal fluid, meninges, skull, and scalp. Furthermore, the distortion of the scalp potential distribution is increased by the ear and eyeholes, which represent shunt paths for intracranial currents [1,2]. As a result, the distribution of the scalp potential shows a low spatial resolution not allowing a reliable localization of the cortical generators of the event-related potentials. Moreover, the variations of electrical reference may enhance or attenuate the spatial components of the potential distribution over the scalp acting as a spatial filter of the cortical generators [2]. For these reasons, the addition of more electrodes is not sufficient per se to improve the spatial information content of an EEG record significantly [1].

3.2 THE HIGH-RESOLUTION ELECTROENCEPHALOGRAPHY

High-resolution EEG technologies have been developed to enhance the spatial information content of EEG activity [1,3]. These technologies consist essentially of high spatial sampling (with 64–128 channels) and surface Laplacian (SL) [4] or spatial deconvolution (SD) estimations [5]. The estimation of the SL of the potential needs the modeling of the scalp surface, while the SD estimation is based on the construction of a multicompartment head volume conductor for simulating cortex, dura mater, skull and scalp surfaces. Most recently, the developed high-resolution EEG enhancement technologies use realistic MRI-constructed subject's head models [5,6]. SL is computed by a spline Laplacian estimator, and SD by a linear inverse estimation method based on boundary-element (BEM) mathematics. The important point of high-resolution EEG technologies is the availability of an accurate model of the head as a volume conductor to be used with advanced computational techniques such as SL or SD. However, appropriate techniques have to be used in order to register the electrode positions on the scalp model. Several authors have shown that it is possible to improve the spatial resolution of EEG by using sophisticated computational algorithms and detailed geometrical models of the head as a volume conductor with the help of the MRI data [1,3,7–10].

3.3   THE SEARCH FOR THE CORTICAL SOURCES

However, the ultimate goal of any EEG or MEG recording is to supply useful information about the brain activity of a subject during a particular task. In order to obtain such information we have to start from these EEG or MEG recordings to arrive at an estimate of cortical activity, by using a body of mathematical techniques known as inverse procedures. Examples of these inverse procedures are the dipole localization, the distributed source and the deblurring or cortical imaging techniques [1,8,11]. Mathematical models for the head as a volume conductor as well as for neural sources are employed by linear and nonlinear minimization procedures to localize putative sources of EEG data. Several studies have indicated the adequacy of the equivalent current dipole as a model for cortical sources [1,2], while the importance of realistic geometry head volume conductor models for the localization of cortical activity was stressed recently [1,8,9,12]. The results of previous intracranial EEG studies support the idea that high-resolution EEG techniques (including head/source models and proper regularization inverse procedures) might model with an acceptable approximation the strengths and the extension of cortical sources of surface EEG data, at least in certain conditions [5,6,10,13].

3.4   HEAD AND CORTICAL MODELS

The estimation of the cortical activity from EEG scalp recordings, based on the solution of the linear inverse problem (showed in the next paragraph) requires the use of electrical models of the structures of the head, interposed between the sensors (electrodes on the scalp) and the electrical sources (the neurons in the cortex). Realistic head models can be reconstructed from T1-weighted MRIs of each experimental subject and used to improve such estimation. In the data analysis performed in this book, scalp, skull, and dura mater compartments were segmented from MRIs and tessellated with about 5000 triangles for each surface (see Fig. 3.1). A source model of the cortex was built using the following procedure: (1) The cortex compartment was segmented from MRIs and tessellated to obtain a fine mesh with about 100,000 triangles; (2) A coarser mesh was obtained by resampling the fine mesh down to about 5000 triangles (this was done by preserving the general features of the neocortical envelope, especially in correspondence of precentral and postcentral gyri and frontal mesial area); (3) An orthogonal unitary equivalent current dipole was placed in each node of the tessellated surface, parallel to the vector sum of the normals to the surrounding triangles. Such a model was used to define the dole source configuration that is the result of the estimation procedure.

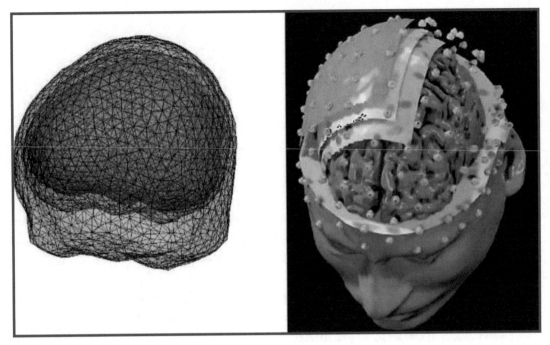

FIGURE 3.1: Realistic models of scalp, skull, dura mater and cortex, tessellated from MRI images of the experimental subject. The electrodes were superimposed to the model by means of a stereophotogrammetric method with a precision <1 mm.

3.4.1   Regions of Interest (ROI)

Thousands of dipoles are used to describe adequately the complex folding of the cortical surface from a mathematical point of view. However, particular regions of the cortex present similar features in terms of the cytoarchitecture (basically, the appearance of the cortex under the light microscope). Then, it could be convenient to use such particular regions in order to estimate the cortical activity, in order to decrease the number of the waveforms to be treated with the connectivity estimators described in the previous chapters. Korbinian Brodmann (1868–1918) was an anatomist who divided the cerebral cortex into numbered subdivisions on the basis of cell arrangements, types, and staining properties (Brodmann classified brain regions based on their cytoarchitecture). In some instances there is a clear link between the microscopic appearance of a region and its function. For example, the "stripe" of the striate cortex delineates the first main cortical area of the visual system (today this area is usually referred to as V1, Brodmann called it area 17). However, it is important to remember that Brodmann's areas (BAs) were identified purely based on visual appearance, which is not necessarily related to function. Figure 3.2 presents the original Brodmann maps. As described in the next pages, several BAs are important in the context of the motor and cognitive tasks.

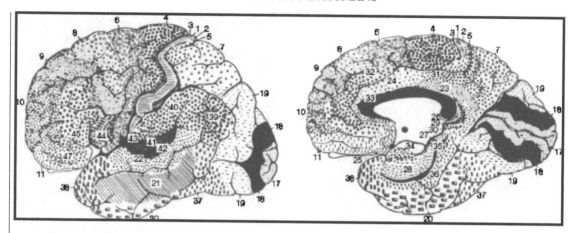

FIGURE 3.2: Brodmann's original cytoarchitectonic maps of the human brain with a ventral view of the brain on the left and a medial view of the cortical areas on the medial wall of the brain on the right.

It is useful to estimate the cortical activity for the cortical patches that have the same cytoarchitectonic properties, i.e., BAs, instead of using data from any single dipole.

The functional connectivity will then be a computing device between the estimated data for the ROIs, depicted along BAs identified on individual cortical model. This strategy uses *a priori* information according to the role of the BAs in the brain functions. The presented technique could also be applied by drawing the ROIs around the cortical estimated peaks of the power spectra activity in the different frequency bands with a post hoc procedure [14]. In this book, the ROIs depicted on the base of the BAs were employed to allow comparison of the functional connectivity patterns elicited by the same experimental behavior across subjects.

The present analysis was based on a cortical source model, namely the source space where the cortical activity generated was identified. Hence, it may be argued that because only cortical sources are modeled, if a deep source is active, then the source reconstruction (and likely the connectivity estimates) could fail. In this context, a widely accepted notion that the main sources for the scalp-recorded EEG signals is derived from the cortex, whereas the thalamus and the basal ganglia can hardly produce appreciable contribution to the scalp EEG should be taken into account [1]. However, even if a subcortical neural source contributed markedly to scalp-recorded EEG, this deep contribution would be distributed over the source space lying on the cortical surface by the employed model. This phenomenon would result in an increase in the low spatial frequency component in the recorded EEG.

3.5 DISTRIBUTED SOURCES ESTIMATE

Accurate estimates of the cortical current density could be obtained by using adequately detailed geometrical reconstruction of the main compartments lying between the cortical generator sources and the EEG or MEG sensors. The estimate of the cortical current density from noninvasive EEG and/or MEG data can be obtained by solving a linear problem. In this problem, the cortical sources to be estimated are related to the noninvasive measurements by means of a transfer matrix (lead field matrix) that mimics the effects of the volume conductor [15,16]. In mathematical terms, the relationship between the modeled sources $\mathbf{x}$, the lead field matrix $\mathbf{A}$, the EEG/ MEG measurements $\mathbf{b}$ and the noise $\mathbf{n}$ can be written as

$$\mathbf{Ax} = \mathbf{b} + \mathbf{n} \qquad (3.1)$$

The solution of this linear system provides an estimation of the dipole source configuration $\mathbf{x}$ that generates the measured EEG potential distribution $\mathbf{b}$. The system includes the measurement noise $\mathbf{n}$, which is assumed to be normally distributed. $\mathbf{A}$ is the lead field or the forward transmission matrix, whose jth column describes the potential distribution generated on the scalp electrodes by the jth unitary dipole. The current density solution vector $\boldsymbol{\xi}$ was obtained as [16]

$$\boldsymbol{\xi} = \arg\min_x(||\mathbf{Ax} - \mathbf{b}||_M^2 + \lambda^2||\mathbf{x}||_N^2) \qquad (3.2)$$

where $\mathbf{M}$ and $\mathbf{N}$ are the matrices associated with the metrics with the data and of the source space, respectively, λ is the regularization parameter and $||\mathbf{x}||_M$ represents the $\mathbf{M}$ norm of the vector $\mathbf{x}$. The solution of Eq. (3.2) is given in the inverse operator $\mathbf{G}$:

$$\boldsymbol{\xi} = \mathbf{Gb}, \qquad \mathbf{G} = \mathbf{N}^{-1}\mathbf{A}'(\mathbf{AN}^{-1}\mathbf{A}' + \lambda\mathbf{M}^{-1})^{-1} \qquad (3.3)$$

An optimal regularization of this linear system was obtained by the L-curve approach [17,18]. As a metric in the data space we used the identity matrix, whereas as a norm in the source space we use the following metric:

$$(\mathbf{N}^{-1})_{ii} = ||\mathbf{A}_i||^{-2} \qquad (3.4)$$

where $(\mathbf{N}^{-1})_{ii}$ is the ith element of the inverse of the diagonal matrix $\mathbf{N}$ and all the other matrix elements N_{ij} are set to 0. The L_2 norm of the ith column of the lead field matrix $\mathbf{A}$ is denoted by $||\mathbf{A}_i||$.

3.5.1   Cortical Estimated Waveforms

Using the relations described earlier, an estimate of the signed magnitude of the dipolar moment for each one of the 5000 cortical dipoles was obtained for each time point. As the orientation of the dipole was defined to be perpendicular to the local cortical surface in the head model, the estimation process returned a scalar vector field. To obtain the cortical current waveforms for all the time points, we used a unique "quasioptimal" regularization λ value for all the analyzed EEG potential distributions. The quasioptimal regularization value was computed as an average of the several λ values obtained by solving the linear inverse problem for a series of EEG potential distributions.

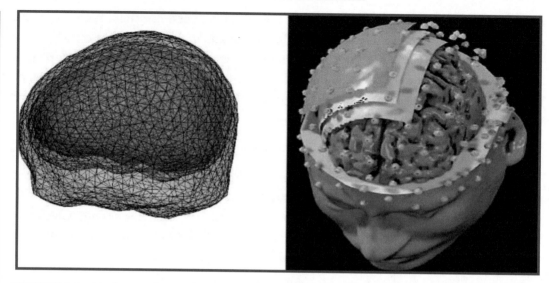

FIGURE 3.3: (*Left*) map of potential recorded on the scalp by means of 58 electrodes. (*Right*) Estimation of the cortical current density obtained from the potentials on the scalp by means of the linear inverse procedure with the use of realistic head models.

These distributions are characterized by an average Global field power (GFP) with respect to the higher and lower GFP values obtained from all the recorded waveforms. An example of the spatial resolution obtained by means of the linear inverse estimation is shown in Fig. 3.3. The instantaneous average of the dipole's signed magnitude belonging to a particular ROI generates the representative time value of the cortical activity in that given ROI. By iterating this procedure on all the time instants of the gathered ERP, the cortical ROI current density waveforms were obtained and they could be taken as representative of the average activity of the ROI, during the task performed by the experimental subjects.

REFERENCES

[1]   P.L. Nunez, *Neocortical Dynamics and Human EEG Rhythms*. New York: Oxford University Press; 1995.

[2]   P. Nunez, *Electric Fields of the Brain*. New York: Oxford University Press; 1981.

[3]   A. Gevins, P. Brickett, B. Costales, J. Le, and B. Reutter, "Beyond topographic mapping: towards functional–anatomical imaging with 124-channel EEG and 3-D MRIs," *Brain Topogr.*, vol. 1, pp. 53–64, 1990, doi:10.1007/BF01128862.

[4]   P.L. Nunez, R.B. Silberstein, P.J. Cadiush, J. Wijesinghe, A.F. Westdorp, and R. Srinivasan, "A theoretical and experimental study of high resolution EEG based on surface Laplacians and cortical imaging," *Electroenceph. Clin. Neurophysiol.*, vol. 90, pp. 40–57, 1994, doi:10.1016/0013-4694(94)90112-0.

[5]   J. Le, and A. Gevins, "A method to reduce blur distortion from EEG's using a realistic head model," *IEEE Trans. Biomed. Eng.*, vol. 40, pp. 517–528, 1993, doi:10.1109/10.237671.

[6]   F. Babiloni, C. Babiloni, F. Carducci, L. Fattorini, C. Anello, P. Onorati, and A. Urbano, "High resolution EEG: a new model-dependent spatial deblurring method using a realistically-shaped MR-constructed subject's head model," *Electroenceph. Clin. Neurophysiol.*, vol. 102, pp. 69–80, 1997, doi:10.1016/S0921-884X(96)96508-X.

[7]   F. Babiloni, C. Babiloni, L. Locche, F. Cincotti, P.M. Rossini, and F. Carducci, "High-resolution electroencephalogram: source estimates of Laplacian-transformed somatosensory-evoked potentials using a realistic subject head model constructed from magnetic resonance images," *Med. Biol. Eng. Comput.*, vol. 38, no. 5, pp. 512–519, Sep 2000, doi:10.1007/BF02345746.

[8]   A. Gevins, "Dynamic functional topography of cognitive task," *Brain Topography*, vol. 2, pp. 37–56, 1989b, doi:10.1007/BF01128842.

[9]   A. Gevins, P. Brickett, B. Reutter, and J. Desmond, "Seeing through the skull: advanced EEGs use MRIs to accurately measure cortical activity from the scalp," *Brain Topogr.*, vol. 4, pp. 125–131, 1991, doi:10.1007/BF01132769.

[10] A. Gevins, J. Le, N. Martin, P. Brickett, and J. Desmond, B. Reutter, "High resolution EEG: 124-channel recording, spatial deblurring and MRI integration methods," *Electroenceph. Clin. Neurophysiol.*, vol. 39, pp. 337–358, 1994, doi:10.1016/0013-4694(94)90050-7.

[11] A.M. Dale, and M. Sereno, "Improved localization of cortical activity by combining EEG and MEG with MRI cortical surface reconstruction:a linear approach," *J. Cogn. Neurosci.*, vol. 5, pp. 162–176, 1993.

[12] A. Gevins, J. Le, H. Leong, L.K. McEvoy, and M.E. Smith, "Deblurring," *J. Clin. Neurophysiol.*, vol. 16, no. 3, pp. 204–213, 1999, doi:10.1097/00004691-199905000-00002.

[13] B. He, Y. Wang, and D. Wu, "Estimating cortical potentials from scalp EEG's in a realistically shaped inhomogeneous head model," *IEEE Trans. Biomed. Eng.*, vol. 46, pp. 1264–1268, 1999, doi:10.1109/10.790505.

[14] J. Gross, J. Kujala, M. Hämäläinen, L. Timmermann, A. Schnitzler, and R. Salmelin, "Dynamic imaging of coherent sources: studying neural interactions in the human brain," *Proc. Natl. Acad. Sci. USA*, vol. 98, no. 2, pp. 694–699, 2001, doi:10.1073/pnas.98.2.694.

[15] R.D. Pascual-Marqui, Reply to comments by Hamalainen, Ilmoniemi and Nunez. In ISBET Newsletter N.6, December 1995. Ed: W. Skrandies., 16–28.

[16] R. Grave de Peralta Menendez, and S.L. Gonzalez Andino, Distributed source models: standard solutions and new developments. In: Uhl, C. (Ed.), Analysis of neurophysiological brain functioning. Springer Verlag, pp.176–201, 1999.

[17] P.C. Hansen, "Analysis of discrete ill-posed problems by means of the L-curve," *SIAM Review*, vol. 34, pp. 561–580, 1992a, doi:10.1137/1034115.

[18] P.C. Hansen, "Numerical tools for the analysis and solution of Fredholm integral equations of the first kind," *Inverse Problems*, vol. 8, pp. 849–872, 1992b, doi:10.1088/0266-5611/8/6/005.

CHAPTER 4

Application: Estimation of Connectivity from Movement-Related Potentials

The estimation of connectivity patterns by using the direct transfer function (DTF) and structural equation model (SEM) on high-resolution electroencephalographic (EEG) recordings has been applied to analyze a simple movement task. In particular, we considered the right hand finger tapping movement, externally paced by a visual stimulus. This task was chosen as it has been very well studied in literature with different brain imaging techniques like EEG or functional magnetic resonance imaging [1–3].

4.1 SUBJECTS AND EXPERIMENTAL DESIGN

Three right-handed healthy subjects (age 23.3 ± 0.58 years, one male and two females) participated in the study after the informed consent was obtained according to University of Illinois at Chicago/Institutional Review Board (UIC/IRB). Subjects were seated comfortably in an armchair with both arms relaxed and resting on pillows and they were requested to perform fast repetitive right finger movements cued by visual stimuli. Ten to fifteen blocks of 2 Hz thumb oppositions were recorded, with each 30 s blocks of finger movement and rest. During motor task, subjects were instructed to avoid eye blinks, swallowing, or any movement other than the required finger movements.

4.2 EEG RECORDINGS

Event-related potential (ERP) data were recorded with 96 electrodes; data were recorded with a left ear reference and submitted to the artifact removal processing. Six hundred ERP trials of 600 ms duration were acquired. The A/D sampling rate was 250 Hz. The surface electromyographic (EMG) activity of the muscle was also recorded. The onset of the EMG response served as zero time. All data were visually inspected, and trials containing artifacts were rejected. Semiautomatic supervised threshold criteria were used for the rejection of trials contaminated by ocular and EMG artifacts, as described in detail elsewhere [4]. After the EEG recording, the electrode positions were digitized using a three-dimensional (3D) localization device with respect to the anatomic landmarks of the head (nasion and two preauricular points). The analysis period for the potentials time locked to the movement execution was set from 300 ms before to 300 ms after the EMG trigger (0 time); the ERP time course was divided in two phases relative to the EMG onset; the first period, labeled as "PRE," marked the 300 ms before the EMG onset and was intended as a generic preparation period; the second labeled as "POST," lasted up to 300 ms after the EMG onset and was intended to signal the arrival of the movements somatosensory feedback. We kept the same PRE and POST nomenclature for the signals estimated at the cortical level.

4.2.1   Selection of Regions of Interest (ROIs)

Several cortical regions of interest (ROIs) were drawn on the computer-based cortical reconstruction of the individual head models (three subjects). The ROIs representing the left and right primary somatosensory (S1) areas included the Brodmann areas (BA) 3, 2, and 1, while the ROIs representing the left and right primary motor (MI) included the BA 4. The ROIs representing the supplementary motor area (SMA) were obtained from the cortical voxels belonging to the BA 6. We further separated the proper and anterior SMA indicated with BA 6P and 6A, respectively. Furthermore, ROIs from the right and the left parietal areas including the BA 5, 7, and the occipital areas (BA 19) were also considered. In the frontal regions, the BA 46, 8, and 9 were also selected. A selection of the gathered ERPs related to the visual paced finger tapping task from one of the subjects is depicted in the left part of Fig. 4.1. The waveforms are relative to the signals gathered from the standard electrode leads of the augmented 10–20 system, represented on the realistic scalp reconstruction of the subject, and represented the average of the artifact-free trials. By means of the linear inverse procedure, the estimation of the current density waveforms in the ROIs of interest was then performed, according to the Eqs. (3.1)–(3.4).

The instantaneous average of the signed magnitude of all the dipoles belonging to a particular ROI was used to estimate the average cortical activity in that ROI, during the entire interval of the experimental task. These waveforms could then be subjected to the MVAR modeling in order to estimate the connectivity pattern between ROIs, by taking into account the time varying increase or decrease of the power spectra in the frequency bands of interest.

The estimated current density waveforms for the same subject are represented in some selected ROIs in the right part of the Fig. 4.1. Relevant cortical activity was observed to be different from baseline in the left ROIs representing parietal (BA 5), premotor (BA 6A), sensorimotor (BA 3, 2, 1, BA 4), and prefrontal (BA 8) cortical areas, whereas similar statistical data were obtained for the ROIs located in the right hemisphere in premotor (BA 6A) and prefrontal (BA 8) cortical areas.

4.2.2   Definition of the *A Priori* Model for the Effective Connectivity

From the results obtained by the simulation study described in the previous chapter, the performances seem to be rather good for a correct assumption of the hypothesized model, but they decrease when even a simple error is made, depending on the error type. This degradation of the performances seems to indicate that detailed cortical areas cannot be used in the first step of the network modeling.

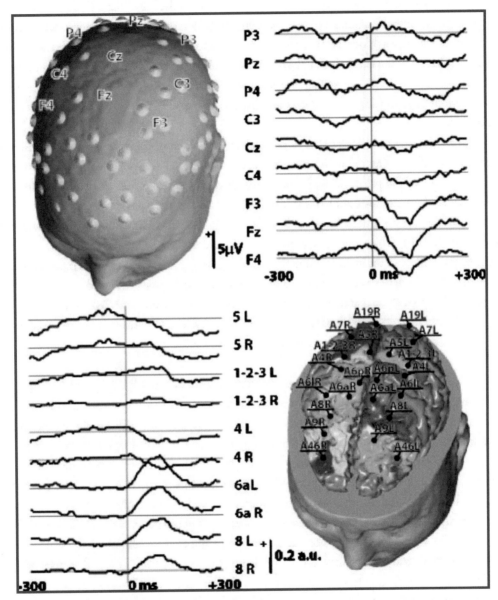

FIGURE 4.1: (*Left*) a selection of the ERPs gathered from the standard electrode leads of the augmented 10–20 system, represented on the realistic geometry scalp reconstruction of the subject (average of the artifact free trials). The onset of the electromyographic (EMG) signal for the start of the movement of the right finger is at the 300 ms from the beginning of the ERP trial. (*Right*) the estimated (via the solution of the linear inverse problem) current density waveforms, represents some selected ROIs on the realistic cortex reconstruction of the same subject. Each ROI is represented in a different grey level. (Published with permission from [5].)

By using a coarse model of the cortical network to be fitted on the EEG data, there is an increase of the statistical power and a decrease of the possibility to generate an error in a single arc link. In the present human study, such observation was taken into account by selecting a coarse model for the brain areas subserving the task being analyzed. This simplified model does not take complete account of all the possible regions engaged in the task, and all the possible connections between them. Elaborate models, permitting also cyclical connections between regions can become computationally unstable [6].

Our model of interactions between cortical areas is based of previous results on similar tasks obtained with different brain imaging methods. It is sufficient to address some key questions regarding the influence of the premotor and motor areas toward the prefrontal cortical areas during the task analyzed. The anatomical model employed is based on the principal cortical areas recognized as active during this simple task in these studies. Cortical areas used in this human study included the prefrontal areas (PF), which includes large BAs 8, 9, and 46; the premotor areas (PM), including the BA 6, the sensorimotor areas (SM) including the BAs 4, 3, 2, and 1, and the parietal areas (P), generated by the union of the BAs 5 and 7. The model employed *a priori* knowledge about the flow of connections between these macroareas, as derived from neuroanatomy and fMRI studies. In particular, information flow was hypothesized to exist from the parietal (P) areas toward the sensorimotor (SM), the premotor (PM) and the prefrontal (PF) ones [1,2,7].

4.2.3   Statistical Evaluation of Connectivity Measurements by SEM and DTF

As described before, the statistical significance of the connectivity pattern estimated with SEM technique was assured by the fact that in the context of the multivariate normally distributed variables, the minimum of the maximum likelihood function F_{ML}, multiplied by $(N-1)$, follows a χ^2 distribution with $(p(p+1)/2) - t$ degrees of freedom, where t is the number of parameters to be estimated and p is the total number of observed variables (endogenous + exogenous). The χ^2 statistic test can then be used to infer statistical significance of the structural equation model obtained.

The situation is different for all areas of the statistical significance of the DTF measurements, as the DTF function has a highly nonlinear relation to the time series data from which they are derived, and the distribution of their estimators is not well established. This makes tests of significance difficult to perform, unless the tests based on this empirical distribution can be performed. A possible solution to this problem was proposed in Kaminski et al. [8]. It consists of the use of a surrogate data technique [9], in which an empirical distribution for random fluctuations of a given estimated quantity is generated by estimating the same quantity from several realizations of surrogate datasets, in which deterministic interdependency between variables were removed. In order to ensure that all features of each dataset are as similar as possible to the original dataset, with the exception of channel coupling, the very same data are used; any time-locked coupling between channels is disrupted by shuffling phases of the original multivariate signal. As the EEG signal had been divided into single trials, each surrogate dataset was built up by scrambling the order of epochs, using different sequences for each channel. The set properties of univariate surrogate signals are not influenced by this shuffling procedure, since only the epoch order is varied. Moreover, since no shuffling was performed between single samples and the temporal correlation, the spectral features, of univariate signals are the same for original and surrogate dataset, thus allowing the estimate of different distributions of DTF fluctuations for each frequency band. One thousand surrogate datasets were generated as described above, and DTF spectra were estimated from each dataset. For each channel pair and for each frequency bin, the 99 percentile was computed and subsequently considered as a significance threshold.

4.2.4   Connectivity Pattern Representation

The connectivity patterns are represented by arrows pointing from one cortical area ("the source") toward another one ("the target"). The grey level and size of the arrow code the strength of the functional connectivity estimated between the source and the target. The bigger and lighter the arrow, the stronger the connection. Only the cortical connections statistically significant at $p < 0.01$ are represented, according to the thresholds obtained as previously described.

The connectivity patterns in the different frequency bands (theta, 4–8 Hz; alpha, 8–12 Hz; beta, 13–30 Hz; gamma, 30–40 Hz), and between the different cortical regions were summarized by using indices representing the total flow from and toward the selected cortical area. The total inflow in a particular cortical region was defined as the sum of the statistically significant connections from all the other cortical regions toward the selected area. A sphere centered on the cortical region, whose radius is linearly related to the magnitude of all the incoming statistically significant links from the other regions, represents the total inflow for each ROI. Inflow information is also coded through a grey level scale. This information depicts each ROI as the target of functional connections from the other ROIs. The same conventions were used to represent the total outflow from a cortical region, generated by the sum of all the statistical significant links.

4.3   RESULTS

The results of the application of the SEM method for the estimation of the connectivity on the event-related potential recordings is depicted in Fig. 4.2, which shows the statistically significant cortical connectivity patterns obtained for the period preceding the movement onset in the subject #1, in the alpha frequency band. Each pattern that connects one cortical area ("the source") to another one ("the target") is represented by arrows. The grey level and sizes of the arrows code the strength level of the functional connectivity observed between ROIs. The labels indicate the names of the ROIs employed. Note that the connectivity pattern during the period preceding the movement in the alpha band involves mainly the parietal left ROI (Pl) coincident with the BAs 5 and 7, functionally connected with the left and right premotor cortical ROIs (PMl and PMr), the left sensorimotor area

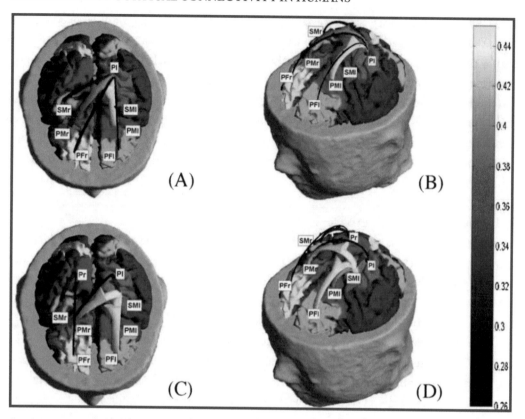

FIGURE 4.2: (A–D) The cortical connectivity pattern obtained for the period preceding and following the movement onset in the subject, in the alpha (8–12 Hz) frequency band. The realistic head model and cortical envelope of the subject analyzed obtained from sequential MRIs is used to display the connectivity pattern. Such a pattern is represented with arrows that move from one cortical area toward another. The grey level and sizes of arrows code the strength level of the functional connectivity observed between ROIs. The labels are relative to the name of the ROIs employed. (A–B) Connectivity patterns obtained from ERP data before the onset of the right finger movement (electromyographic onset; EMG), from above (*left*) and from the left of the head (*right*). (C–D) Connectivity patterns obtained after the EMG onset. (Published with permission from [10].)

(SMl), and both the prefrontal ROIs (PFl and PFr). The stronger functional connections are relative to the link between the left parietal and the premotor areas of both cerebral hemispheres. After the preparation and the beginning of the finger movement POST period changes in the connectivity pattern can be noted. In particular, the origin of the functional connectivity links is positioned in the sensorimotor left cortical areas (SMl). From there, functional links are established with prefrontal left (PFl), and both the premotor areas (PMl. PMr). A functional link emerged in this condition connecting the right parietal area (Pr) with the right sensorimotor area (SMr). The left parietal area

(Pl), much active in the previous condition, was linked instead with the left sensorimotor (SMl) and right premotor (PMr) cortical areas.

Figure 4.2 shows the cortical connectivity patterns obtained for the period preceding the movement onset in the subject #1, in the alpha frequency band. Each pattern is represented by arrows that connect one cortical area to another. The grey level and sizes of arrows code the strength level of the functional connectivity observed between ROIs. The labels indicate the names of the ROIs employed. Note that the connectivity pattern during the period preceding the movement in the alpha band involves mainly the parietal left ROI (Pl) coincident with BAs 5 and 7, which is functionally connected with the left and right premotor cortical ROIs (PMl and PMr), the left sensorimotor area (SMl), and both the prefrontal ROIs (PFl and PFr). The stronger functional connections are relative to the link between the left parietal and the premotor areas of both cerebral hemispheres. After the preparation and the beginning of the finger movement, in the POST period changes in the connectivity pattern can be noted.

In particular, the origin of the functional connectivity links is positioned in the sensorimotor left cortical areas (SMl). From there, functional links are established with prefrontal left (PFl), both the premotor areas (PMl. PMr). A functional link emerged in this condition connecting the right parietal area (Pr) with the right sensorimotor area (SMr). The left parietal area (Pl) much active in the previous condition was linked instead with the left sensorimotor (SMl) and right premotor (PMr) cortical areas.

The connectivity estimations performed by DTF on the movement-related potentials were first analyzed from a statistical point of view via the shuffling procedure described. The order of the MVAR model used for each DTF estimation had to be determined for each subject and in each temporal interval of the cortical waveforms segmentation (PRE and POST interval). The Akaike information criterion (AIC) procedure was used and it returned an optimal order between 6 and 7 for all the subjects, in both PRE and POST intervals. On such cortical waveforms, the DTF computational procedure described in the Methods section was applied. Figure 4.3 shows the cortical connectivity patterns obtained for the period preceding and following the movement onset in the subject #1. Here, we present the results obtained for the connectivity pattern in the alpha band (8–12 Hz), since the ERP data related to the movement preparation and execution are particularly responsive in such frequency interval (for a review, see [12]). The task-related pattern of cortical connectivity was obtained by calculating the DTF between the cortical current density waveforms estimated in each ROI depicted on the realistic cortex model. The connectivity patterns between the ROIs have been represented by arrows pointing from one cortical area toward another one. The grey level and size of the arrow code the strength of the functional connectivity estimated between the source and the target ROI. Labels indicate the ROIs involved in the estimated connectivity pattern. Only the cortical connections statistically significant at $p < 0.01$ are represented, according to the thresholds obtained by the shuffling procedure described earlier. It can be noted that the

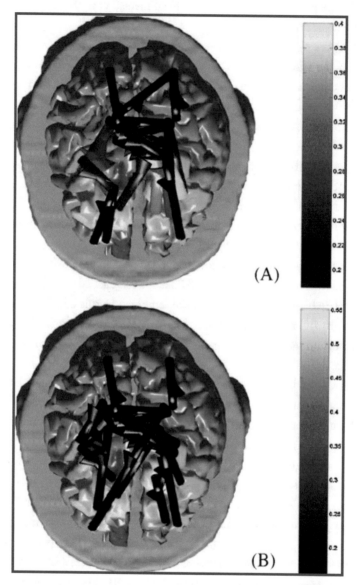

FIGURE 4.3: Cortical connectivity patterns obtained by the DTF method, for the period preceding and following the movement onset, in the alpha (8–12 Hz) frequency band. The patterns are shown on the realistic head model and cortical envelope of the subject analyzed, obtained from sequential MRIs. Functional connections are represented with arrows, moving from a cortical area toward another one. The grey level and sizes of the arrows code the strengths of the connections. (A) Connectivity pattern obtained from ERP data before the onset of the right finger movement (electromyographic onset; EMG). (B) Connectivity patterns obtained after the EMG onset. Same conventions as described earlier. (Published with permission from [11].)

connectivity patterns during the period preceding and following the movement in the alpha band involves bilaterally the parietal and sensorimotor ROIs, which are also functionally connected with the premotor cortical ROIs. A minor involvement of the prefrontal ROIs is also observed. The stronger functional connections are relative to the link between the premotor and prefrontal areas of both cerebral hemispheres. After the preparation and the beginning of the finger movement, in the POST period, slight changes in the connectivity patterns can be noted.

4.4   DISCUSSION

In the case in which the SEM methodology was applied on the recorded high-resolution EEG data the model of interactions between cortical areas is based on previous results on similar tasks obtained with different brain imaging methods. Such a model is sufficient to address some key questions regarding the influence of the premotor and motor areas toward the prefrontal cortical areas during the task analyzed. The finger tapping data analyzed here present a high SNR and a large number of trials, resulting in an extended record of ERP data. Hence, the present simulation results suggest the optimal performance of the SEM method as applied to the human ERP potentials. The connectivity pattern estimated via SEM reveals the potentiality of the employed methodology including the use of high-resolution EEG recordings, the generation of a realistic head model by using sequential MRIs, and the estimation of the cortical activity with the solution of linear inverse problem. With this methodology, it would be possible not only to detect which of the cortical areas are activate during a particular (motor) task but also how these areas are effectively interconnected in subserving that given task. In particular, the influence of the parietal area has been observed toward the premotor cortical areas during the task preparation, consistent with the role that the parietal areas have in the engagement of attentive resources as well as temporization, as assessed by several electrophysiological studies on primate or hemodynamical studies on humans [13] It is interesting to note the cortical areas behaving as the most relevant origin of functional links, occurring when the somatosensory reafferences arrive from the periphery to the cortex. In fact, the left sensorimotor area becomes very active with respect to the left parietal one, which, in turn, is mainly engaged in the time period preceding the finger movement. Connections between the sensorimotor area and the premotor and left prefrontal areas are appropriate to distribute the information related to the movement of the finger to the higher functional regions (prefrontal and premotor).

From a physiological point of view, the results obtained by estimating the connectivity patterns with the DTF are consistent and integrate those already present in literature on simple finger movements, as they have been obtained with neuroelectric measurements. A study employing ERP measurements from scalp electrodes and the assessment of connectivity with the non directional coherence methods has underlined the role of the primary sensorimotor and supplementary motor areas in the processing of the movements [1]. The connectivity patterns depicted in the premotor and prefrontal ROIs here analyzed are in agreement with earlier hypotheses formulated in literature

[14–16]. The aforementioned studies have suggested that the dorsolateral and the ventral premotor cortices are specifically activated by movements guided by sensory information as opposed to movements carried out with no sensory control. The activity noted in the parietal areas (BA 5) in the present study, could be associated with the role that this area has in the somatosensory–motor integration underlying movement performing. In fact, it has been hypothesized that this area could be regarded as a higher order somatosensory zone devoted to the analysis of proprioceptive information from joints for the appropriate motor control [17].

REFERENCES

[1] C. Gerloff, J. Richard, J. Hadley, A. E. Schulman, M. Honda, and M. Hallett, "Functional coupling and regional activation of human cortical motor areas during simple, internally paced and externally paced finger movements," *Brain*, vol. 121, no. Pt 8, pp. 1513–1531, 1998, doi:10.1093/brain/121.8.1513.

[2] A. S. Gevins, B. A. Cutillo, S. L. Bressler, N. H. Morgan, R. M. White, J. Illes, and D.S. Greer, "Event-related covariances during a bimanual visuomotor task. II. Preparation and feedback," *Electroencephalogr. Clin. Neurophysiol.*, vol. 74, pp. 147–160, 1989, doi:10.1016/0168-5597(89)90020-8.

[3] L. Jancke, R. Loose, K. Lutz, K. Specht, and N. J. Shah, "Cortical activations during paced finger-tapping applying visual and auditory pacing stimuli," *Brain Res. Cogn. Brain Res.*, vol. 10, no. 1–2, pp. 51–66, 2000, doi:10.1016/S0926-6410(00)00022-7.

[4] D. V. Moretti, F. Babiloni, F. Carducci, F. Cincotti, E. Remondini, P. M. Rossini, S. Salinari, and C. Babiloni, Computerized processing of EEG-EOG-EMG artifacts for multi-centric studies in EEG oscillations and event-related potentials," *Int. J. Psychophysiol.*, vol. 47, no. 3, 199–216, Mar. 2003, doi:10.1016/S0167-8760(02)00153-8.

[5] L. Astolfi, et al., "Assessing cortical functional connectivity by linear inverse estimation and directed transfer function: simulations and application to real data," *Clinical Neurophysiol.*, vol. 116, no. 4, pp. 920–932, Apr 2005, doi:10.1016/j.clinph.2004.10.012.

[6] A. R. McIntosh and F. Gonzalez-Lima, "Structural equation modeling and its application to network analysis in functional brain imaging," *Hum. Brain Mapp.*, vol. 2, pp. 2–22, 1994, doi:10.1002/hbm.460020104.

[7] A. Urbano, C. Babiloni, P. Onorati, and F. Babiloni, "Dynamic functional coupling of high resolution EEG potentials related to unilateral internally triggered one-digit movements," *Electroencephalogr. Clin. Neurophysiol.*, vol. 106, no. 6, pp. 477–487, 1998, doi:10.1016/S0013-4694(97)00150-8.

[8] M. Kaminski, M. Ding, W.A. Truccolo, and S. Bressler, "Evaluating causal relations in neural systems: Granger causality, directed transfer function and statistical assessment of significance," *Biol. Cybern.*, vol. 85, pp. 145–157, 2001, doi:10.1007/s004220000235.

[9] J. Theiler, S. Eubank, A. Longtin, B. Galdrikian, and J. D. Farmer, "Testing for nonlinearity in time series: the method of surrogate data," *Physica D*, vol. 58, pp. 77–94, 1992, doi:10.1016/0167-2789(92)90102-S.

[10] L. Astolfi, et al., "Estimation of the cortical connectivity by high resolution eeg and structural equation modeling: simulations and application to finger tapping data," *IEEE Trans. Biomed. Eng.*, vol. 52, no. 5, pp. 757–768, May 2005, doi:10.1109/TBME.2005.845371.

[11] L. Astolfi, et al., "Estimation of the effective and functional human cortical connectivity with structural equation modeling and directed transfer function applied on high resolution EEG," *Magn. Reson. Imaging*, vol. 22, no. 10, pp. 1457–1470, Dec. 2004, doi:10.1016/j.mri.2004.10.006.

[12] G. Pfurtscheller and F. H. Lopes da Silva, "Event-related EEG/MEG synchronization and desynchronization: basic principles," *Clin. Neurophysiol.*, vol. 110, no. 11, pp. 1842–1857, Nov 1999, doi:10.1016/S1388-2457(99)00141-8.

[13] J. C. Culham and N. G. Kanwisher, "Neuroimaging of cognitive functions in human parietal cortex," *Curr. Opin. Neurobiol.*, vol. 11, no. 2, pp. 157–163, 2001, doi:10.1016/S0959-4388(00)00191-4.

[14] J. C. Rothwell, P. D. Thompson, B. L. Day, S. Boyd, and C. D. Marsden, "Stimulation of the human motor cortex through the scalp," *Exp. Physiol.*, vol. 76, no. 2, pp. 159–200, 1991.

[15] K. Sekihara and B. Scholz, "Generalized Wiener estimation of three-dimensional current distribution from biomagnetic measurements," *IEEE Trans. Biomed. Eng.*, vol. 43, no. 3, pp. 281–291, 1996, doi:10.1109/10.486285.

[16] J. Classen, C. Gerloff, M. Honda, and M. Hallet, "Integrative visuomotor behavior is associated with interregionally coherent oscillation in the human brain," *J. Neurophysiol.*, vol. 3, pp. 1567–1573, 1998.

[17] G. Rizzolatti, G. Luppino, and M. Matelli, "The organization of the cortical motor system: new concepts," *Electroencephalogr. Clin. Neurophysiol.*, vol. 106, no. 4, pp. 283–296, 1998, doi:10.1016/S0013-4694(98)00022-4.

CHAPTER 5

Application to High-Resolution EEG Recordings in a Cognitive Task (Stroop Test)

5.1   SUBJECTS AND EXPERIMENTAL DESIGN

High-density electroencephalography (EEG) recordings were performed on a group of five normal subjects. We present here the connectivity results related to one representative subject to illustrate the potential of the methods investigated. Subjects were seated in a comfortable chair in a quiet room, which was connected to the adjacent equipment room by intercom. They viewed a screen where the name of a color (e.g., "red") was printed in the same color (e.g., in red ink, congruent condition) or in a different color (e.g., in blue ink, incongruent condition), as shown in Fig. 5.1. Blocks of congruent or incongruent words alternated with blocks of neutral words (not color names). There were 256 trials in 16 blocks (four color congruent, eight neutral, four color incongruent) of 16 trials, with a variable intertrial interval averaging 2000 ms between trial onsets. Within the congruent blocks, half of the words were neutral so as to prevent the development of word reading strategies in the congruent blocks. The trial began with the presentation of a word for 1500 ms, followed by a fixation cross for an average of 500 ms. Each trial consisted of one word presented in one of four ink colors (red, yellow, green, and blue), with each color occurring equally often with each word type (congruent, neutral, incongruent). Subjects were asked to press one of the four buttons that corresponded to the color of the ink in which the word was presented. Data from 0 to 450 ms poststimulus were analyzed.

5.2   EEG RECORDINGS

EEG was recorded using well-established methods. A custom-designed Falk Minow cap located 64 scalp locations for EEG and EOG recordings, with the EEG electrodes spaced equidistantly, with the left mastoid serving as a reference for all other sites. Electrode impedances were below 10 kΩ. Amplifier bandpass was 0.1 to 100 Hz with digitization at 250 Hz, and the EEG data were successively digitally filtered at 50 Hz. Electrode positions were digitized using a Zebris 3D localization device with respect to anatomic landmarks (nasion and two preauricular points). ERP data were visually inspected, and trials containing artifacts were rejected. Semiautomatic, supervised threshold criteria were used for the rejection of trials contaminated by ocular and EMG

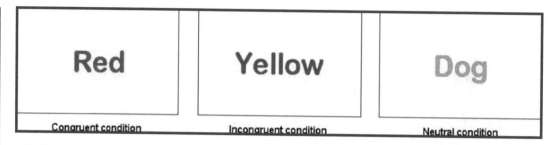

FIGURE 5.1: Scheme of the Stroop task. If the meaning of the word is a color written in the color itself, the condition is congruent (*left*, *word "red" written in red*). If the word means a color and it is written in a different color, the condition is incongruent (*center*, *word "yellow" written in blue*). If the meaning is not a color, the condition is neutral (*right*, *word "dog" written in green*). The subject should pronounce the color in which the word is written (irrespective of the meaning of the word) in the shortest time possible.

artifacts, as described in detail elsewhere [1]. After artifact rejection, ERP signals were adjusted in the baseline.

5.3   RESULTS

After obtaining the solution of the linear inverse problem, the estimation of the current density waveforms on each ROI was obtained as described in the preceding paragraphs. Connectivity estimations were performed by DTF, PDC, and dDTF, and the statistical thresholds were evaluated using the shuffling procedure described earlier. The order of the MVAR model used in each estimation was determined by means of the Akaike information criterion (AIC), which returned an optimal order of 13. Details of the electrode montage are shown on the realistic reconstruction of a subject's scalp in Fig. 5.2. The different ROIs selected are shown in different grey tones on the realistic reconstruction of the subject's left cortical hemisphere (regions of the cortex not of interest are shown in gray). By means of the linear inverse procedure, the estimation of the current density waveforms in each ROI of interest was then performed, according to Eqs. (3.3)–(3.6). The DTF, PDC, and dDTF estimators were applied to the cortical waveforms related to the ROIs of interest.

Figure 5.3 shows the cortical connectivity patterns obtained for the congruent stimuli during the period preceding the subject's answer (0–450 ms after the stimulus presentation) in the representative subject. The results are shown for the beta band (12–29 Hz). The DTF (left), PDC (center), and dDTF (right) methods produced similar results. In particular, functional connections between cortical parietofrontal areas were present in all the estimations performed by the DTF, PDC, and dDTF methods. Moreover, connections involving the cingulate cortex are also clearly visible, as well as those involving prefrontal areas, mainly in the right hemisphere. Functional connections in prefrontal and premotor areas tended to be right-sided, whereas the functional activity in the parietal cortices was generally bilateral.

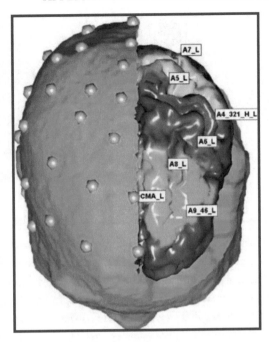

FIGURE 5.2: Over the right hemisphere, the electrode montage (59 electrodes) is shown on the realistic reconstruction of a subject's scalp, obtained from structural MRIs. Over the left hemisphere, the ROIs considered for this study are shown on the realistic reconstruction of the subject's cortex. Each ROI is represented in a different grey level. The ROIs considered are the cingulate motor area (CMA), the Brodmann areas 7 (A7) and 5 (A5), the primary motor area the posterior supplementary motor area (SMAp) and the lateral supplementary motor area (A6_L). (Published with permission from [8].)

Figure 5.4 shows the inflow (first row) and outflow (second row) patterns computed for the same subject in the beta frequency band for the same time period shown in Fig. 5.3. The ROIs that are very active as source or sink (i.e., the source/target of the information flow to/from other ROIs) show results that are generally stable across the different estimators. Values in the left column are related to the inflow and outflow computations obtained with the DTF methods, the central column is related to the flows obtained with the PDC, and the right column to the dDTF. Greater involvement of the right premotor and prefrontal regions is observed across methods.

5.4   DISCUSSION

Although presented here only to demonstrate the capabilities of the estimation procedures with real data, the physiological results shown for a representative subject are consistent with those present in the Stroop literature. The Stroop task is often employed in studies of selective attention and has been found to be sensitive to prefrontal damage. For incongruent stimuli, PET and fMRI studies have shown activation of a network of anterior brain regions. Most studies report

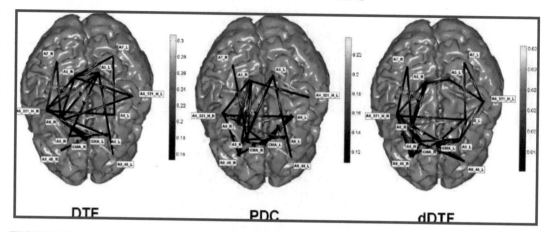

FIGURE 5.3: Cortical connectivity patterns obtained for the period preceding the subject's response during congruent trials in the beta (12–29 Hz) frequency band in a representative subject. The patterns are shown on the realistic head model and cortical envelope of the subject, obtained from sequential MRIs. The brain is seen from above with the left hemisphere represented on the right side. Functional connections are represented by arrows that move from one cortical area toward another. The grey level and sizes of the arrows code the strengths of the connections. The lighter and bigger the arrows are, the stronger the connections. Three connectivity patterns are depicted, estimated in the beta frequency band for the same subject with the DTF (*left*), the PDC (*middle*) and the dDTF (*right*). Only the cortical connections statistically significant at $p < 0.01$ are reported. (Published with permission from [8].)

activation of the anterior cingulate cortex (ACC) and frontal polar cortex, and several authors have hinted at changes in regional cerebral blood flow (rCFB) in posterior cingulate and other posterior regions [2–4].

In the present study, ACC was approximately modeled by the CMA ROI and dorsolateral prefrontal areas by the 9/46 ROI. Connectivity analysis indicated intense, bilateral ACC activity during the task. The number of directed interactions was similar across several frequency bands, but there were differences in the connectivity structure across the congruent–incongruent tasks (not shown here). These results were corroborated by the inflow–outflow analysis that showed agreement between the methodologies used for the derivation of the connectivity patterns. The predominance of outflow from right premotor and prefrontal cortical areas is of interest, and the increased activity in the prefrontal cortical regions is in agreement with previous scalp observations. In an EEG Stroop study, West and Bell [5] reported increased spectral power at medial (F3, F4) and lateral (F7, F8) frontal sites, as well as over parietal regions (P3, P4). They suggested that activation of the parietal cortex resulted from interaction between prefrontal and parietal regions during the suppression of the influence of the irrelevant word meaning. Interaction between parietal and frontal sites has been advocated as an explanation for the activation of posterior areas [3,5,6] Possibly, the minimization

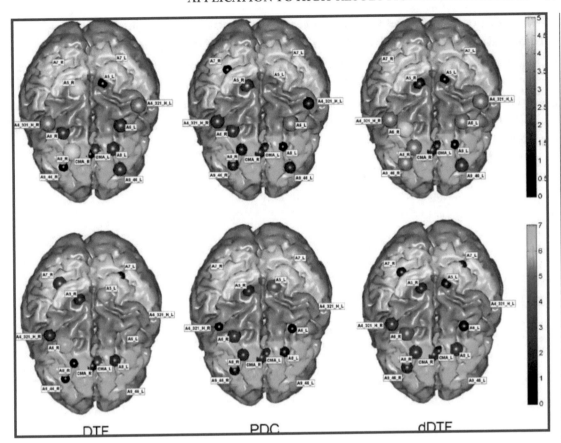

DTF   PDC   dDTF

FIGURE 5.4: Figure shows the inflow (*first row*) and the outflow (*second row*) patterns obtained for the beta frequency band, from each ROI during the congruent trials. The brain is seen from above with the left hemisphere represented on the right side. In the first row, different grey hues summarize the behavior of an ROI in terms of reception of information flow from other ROIs, by adding the values of the links arriving on the particular ROI from all the others. The information is coded with the size and the grey level of a sphere, centered on the particular ROI analyzed. The larger the sphere, the higher the value of inflow or outflow for any given ROI. In the second row, the different grey hues code the outflow of information from a single ROI toward all the others. (Published with permission from [8].)

of the influence of irrelevant word information prompts directed interactions from parietal toward frontal sites.

In a previous coherence study [7], increased coherence between parietal and frontal sites was observed late in the trial. This behaviour is not detectable at the scalp level. With the application of advanced high-resolution EEG methodologies, including realistic cortical modeling, solution of the linear inverse problem, and the application to the computed cortical signals of connectivity pattern estimators, it became observable. These data suggest that cognitive control is implemented by medial and lateral prefrontal cortices that bias processes in regions that have been implicated in high-level perceptual and motor processes [8]. It is also striking that in all the frequency bands and for the five subjects analyzed, the differences between the estimated connectivity patterns with the connectivity estimation methods were negligible (results not shown here). This result suggests that, in practical conditions, constant differences of only a few percent observed in the simulation studies between DTF, PDC, and dDTF estimators are not as significant as other factors such as SNR and recording LENGTH.

REFERENCES

[1] D. V. Moretti, F. Babiloni, F. Carducci, F. Cincotti, E. Remondini, P. M. Rossini, S. Salinari, and C. Babiloni, "Computerized processing of EEG-EOG-EMG artifacts for multi-centric studies in EEG oscillations and event-related potentials," *Int. J. Psychophysiol.*, vol. 47, no. 3, pp. 199–216, Mar 2003, doi:10.1016/S0167-8760(02)00153-8.

[2] L. Astolfi, et al., "A Comparison of different cortical connectivity estimators for high resolution EEG recordings," *Hum. Brain Mapp.*, vol. 28, no. 2, pp. 143–157, Feb 2007, doi:10.1002/hbm.20263.

[3] C. S. Carter, M. Mintun, and J. D. Cohen, "Interference and facilitation effects during selective attention: an H215O PET study of Stroop task performance," *NeuroImage*, vol. 2, pp. 264–272, 1995, doi:10.1006/nimg.1995.1034.

[4] C. J. Bench, C. D. Frith, P. M. Grasby, K. J. Friston, E. Paulesu, R. S. J. Frackowiak, and R. J. Dolan, "Investigations of the functional anatomy of attention using the Stroop test," *Neuropsychology*, vol. 31, pp. 907–922, 1993, doi:10.1016/0028-3932(93)90147-R.

[5] M. P. Milham, M. T. Banich, and V. Barad, "Competition for priority in processing increases prefrontal cortex's involvement in top-down control: an event-related fMRI study of the Stroop task," *Cogn. Brain Res.*, vol. 17, pp. 212–222, 2003, doi:10.1016/S0926-6410(03)00108-3.

[6] R. West and M. A. Bell, "Stroop-color word interference and electroencephalogram activation: evidence for age-related decline of the anterior attention system," *Neuropsychology*, vol. 11, pp. 421–427, 1997, doi:10.1037/0894-4105.11.3.421.

[7] B. Schack, A. C. N. Chen, S. Mescha, and H. Witte, "Instantaneous EEG coherence analysis during the Stroop task," *Clin. Neurophysiol.*, vol. 110, pp. 1410–1426, 1999, doi:10.1016/S1388-2457(99)00111-X.

[8] T. Egner and J. Hirsh, "The neural correlates and functional integration of cognitive control in a Stroop task," *NeuroImage*, vol. 24, pp. 539–547, 2005, doi:10.1016/j.neuroimage.2004.09.007.

CHAPTER 6

Application to Data Related to the Intention of Limb Movements in Normal Subjects and in a Spinal Cord Injured Patient

6.1 SUBJECTS AND EXPERIMENTAL DESIGN

Three healthy subjects and one with a spinal cord injury (SCI) participated in the study. Informed consent was obtained from each subject after explanation of the study, which was approved by the local institutional ethics committee. The SCI was of traumatic etiology and was located at the cervical level (C7); the patient had not suffered a head or brain lesion associated with the trauma leading to the injury. The patient was unable to move his upper and lower limbs. For the electroencephalography (EEG) data acquisition, the subjects were comfortably seated on a reclining chair in an electrically shielded, dimly lit room. They were asked to perform a brisk protrusion of their lips (lip pursing) while they were performing (normal subjects) or attempting (SCI patient) a right foot movement. The task was repeated every 6–7 s, in a self-paced manner, and the 100 single trials recorded were used for the estimation of DTF. A 96-channel EEG system (BrainAmp, Brainproducts GmbH, Germany) was used to record electrical potentials by means of an electrode cap accordingly to an extension of the 10–20 international systems. Structural MRIs of the subjects' head were taken with a Siemens 1.5T Vision Magnetom MR system (Germany).

6.2 ESTIMATED CONNECTIVITY PATTERNS

After the solution of the linear inverse procedure, the estimation of the current density waveforms in the employed ROIs were obtained as previously described. Statistical significance of the cortical connections was obtained by comparing the estimated cortical connectivities with the mean values of the distribution of the random connectivity values between the cortical signals after the deterministic interdependency between these signals were removed. Fig. 6.1 shows the cortical connectivity patterns during the period preceding the onset of the lips movement, and hence related to the preparation of the foot and lips movement in all the three normal subjects examined. Here, we present the results obtained for the connectivity pattern in the gamma band. In the inset, the grey level and size of

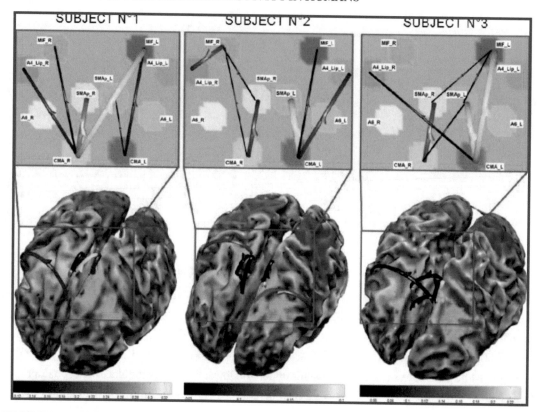

FIGURE 6.1: The cortical connectivity patterns obtained for the period preceding the lips movement in three normal subjects, analyzed in the gamma frequency band. Cortical functional connections are represented with arrows that move from the source cortical area toward the target one. In the inset, the grey level and size of the arrows code the strength level of the connections. (*Bottom*) Connectivity patterns obtained from EEG data represented on the realistic cortical reconstruction of each experimental subject obtained from sequential MRIs, seen from left and above. (*Top*) Details of the connectivity patterns obtained for the central areas. Only the cortical connections statistically significant at $p < 0.01$ are represented. (Published with permission from [1].)

the arrows code the strength level of the connection. In the labels, the names of the ROIs employed are indicated. Only the cortical connections statistically significant at $p < 0.01$ are represented. Note that the connectivity patterns, estimated in the gamma band, presents strong functional connections between the CMA and the premotor areas of both cerebral hemispheres Substantial equivalence of the connectivity patterns estimated for the three normal subjects was found. Hence, the functional directional connections during the preparation of the foot movement are generated in the gamma band from the cingulated areas and spread toward the supplementary motor areas. These patterns have to be compared with those estimated in the SCI patient, during the performance of a similar experimental task.

6.3   CORTICAL CONNECTIVITY PATTERNS IN SPINAL CORD INJURY

The EEG recording and the estimation of the cortical activity and connectivity for the SCI patient during the task was accomplished in the same way already described for the normal subjects. Fig. 6.2 depicts the connectivity pattern in the gamma frequency band before the execution of the lip movement accompanied by the attempt to move the paralyzed limb. It can be noted how this pattern is similar to those generated by the normal subjects during the preparation to the foot movement. Also in this case the connectivity flow is generated from the cingulated areas and spreads to the supplementary motor areas. Similar was the case observed for the connectivity patterns already generated

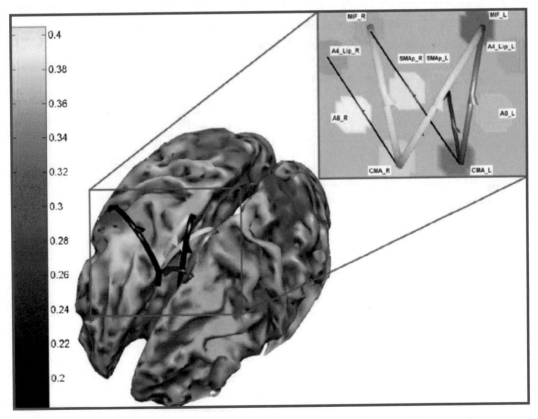

FIGURE 6.2: Cortical connectivity patterns obtained for the period preceding the onset of movement in a SCI patient in the gamma frequency band. Same convention is used as shown in the previous figure. Only the cortical connections statistically significant at $p < 0.01$ are represented. (Published with permission from [1].)

in the other frequency bands examined between the SCI patient and the group of three normal subjects.

6.4   CONCLUSION

The main results provided here highlight the possible existence of a common pattern of cortical connectivity during the execution (normal) or the imagination (SCI patient) of a foot limb movement. The activity noted in the cingulated and supplementary motor areas in the present study is consistent with the role that such cortical areas have in the organization and performance of simple foot movements. This finding, if confirmed in a larger population of normal subjects as well as SCI patients, could open the way for the use of such a feature in a clinical context, for instance, in the brain–computer interface area. It is worth noting that the technology presented here can be applied to retrieve patterns of cortical connectivity during more complex clinically relevant tasks in patients by using noninvasive EEG recordings. Examples include the use of the connectivity pattern study in the analysis of the brain damage caused by a stroke as well as in the analysis of the possible recovery of the brain motor areas during the rehabilitation paths.

REFERENCES

[1] L. Astolfi, et al., "Cortical connectivity patterns during the intention of limb movements in normal and in a spinal cord injured patient estimated with the Directed Transfer Function," *IEEE Eng. Med. Biol. Mag.*, vol. 25, no. 4, pp. 32–38, Jul-Aug 2006.

CHAPTER 7

The Instantaneous Estimation of the Time-Varying Cortical Connectivity by Adaptive Multivariate Estimators

7.1 TIME-VARYING ESTIMATION OF THE CORTICAL CONNECTIVITY

Taken together, the findings of this part of the research indicate that an accurate estimation of the cortical connectivity patterns can be achieved by using realistic models for the head and cortical surfaces, high-resolution EEG recordings, and effective and functional cortical connectivity by using the SEM, DTF, dDTF, and PDC methods, respectively. The simulation findings suggest that under conditions largely met in the ERPs recordings (SNR at least 3 and the EEG recording with 4800 data samples or duration longer than 75 s at 64 Hz), the computation of functional connectivity by all the methods employed can be performed with moderate quantitative errors. The use of high-resolution EEG recordings and the estimation of the cortical waveforms in ROIs through the solution of linear inverse problem allow the evaluation of the functional cortical connectivity patterns related to the task performed. These computational tools (high-resolution EEG, estimation of cortical activity via linear inverse problem, and connectivity methods) can be of interest to assess functional connectivity patterns from noninvasive EEG recordings in humans. However, the classical estimation of these methods requires the stationarity of the signals, and with the estimation of a unique MVAR model on an entire time interval, transient pathways of information transfer remains hidden. This limitation could bias the physiologic interpretation of the results obtained with the connectivity technique employed.

To overcome this limitation, we need to approach a time-varying estimation of cortical connectivity to be performed always within the framework of the Granger causality. This is the aim of this chapter.

As described in Chapter 2, among the multivariate methods, the directed transfer function (DTF) [1,2] and the partial directed coherence [3] are estimators characterizing, at the same time, direction and spectral properties of the interaction between brain signals, and require only one MVAR model to be estimated from all the time series. However, the classical estimation of these methods requires the stationarity of the signals; moreover, with the estimation of a unique MVAR model on an entire time interval, transient pathways of information transfer remains hidden. This limitation could bias the physiologic interpretation of the results obtained with the connectivity technique employed.

To overcome this limitation, different algorithms for the estimation of MVAR with time-dependent coefficients were recently developed. Ding et al. [4] used a short-time windows technique that requires the stationarity of the signal within short-time windows. Moeller et al. [5] proposed an application to MVAR estimation of the extension of the recursive least squares (RLS) algorithm with a forgetting factor. This estimation procedure allows the simultaneous fit of one mean MVAR model to a set of single trials, each one representing a measurement of the same task. In contrast to short-window techniques, the multitrial RLS algorithm does not require the stationarity of the signals, and involves the information of the actual past of the signal, whose influence decreases exponentially with the time distance to the actual samples. The advantages of this estimation technique are an effective computation algorithm and a high adaptation capability. It was demonstrated in [5] that the adaptation capability of the estimation (measured by its adaptation speed and variance) does not depend on the model dimension. Simulations on the efficacy of time-variant Granger causality based on AMVAR computed by RLS algorithm were also provided [6].

In this chapter, we propose the use of the adaptive multivariate approach to define time-varying multivariate estimators based on DTF and PDC, thus making such estimators able to follow rapid changes in the connectivity between cortical areas during an experimental task. Further, the performances of time-varying DTF and PDC were studied by means of simulations performed on the basis of a predefined connectivity scheme linking three cortical areas. Cortical connections between the areas were retrieved by the estimation process under different experimental conditions. The results obtained for the different methods were evaluated by a statistical analysis, with particular attention to the adaptation speed and precision of the pattern retrieved. The simulation study was based on the following questions:

1. What are the performances of the proposed time-varying estimators in retrieving the rapid changes in time of the cortical connectivity pattern?

2. What is the effect of different factors affecting the recordings, such as the signal-to-noise ratio and the amount of trials at disposal?

3. What is the influence of the adaptation constant C on the performances, and what can be a criterion for the choice of its optimum value?

4. Which of the two connectivity estimators applied (i.e., PDC and DTF) is the most effective in reconstructing a connectivity model under the conditions usually met in linear inverse estimations?

Finally, according to the results of the simulation study an application to real data is proposed in order to offer an example of the results that can be obtained by this technique. For this purpose, patterns of cortical connectivity between human brain areas involved in a simple motor task are presented. We applied the time-varying DTF/PDC techniques to the cortical activity estimated in particular regions of interest (ROIs) of the cortex, obtained from high-resolution EEG recordings during the execution of a combined foot–lips movement.

The directed transfer function (DTF) and the partial directed coherence (PDC) estimators have been fully introduced in Chapter 2.

In this chapter, an adaptive formulation of DTF and PDC based on an adaptive MVAR (AMVAR) model is proposed and tested. The time-dependent parameter matrices were estimated by means of RLS algorithm with forgetting factor, as described in [5]. In particular, the RLS algorithm represents a particular variant of the Kalman filter. This recursive estimator for the AMVAR-parameter is characterized by a more universal practicability since it requires less computational effort and it is possible to extend this approach to the presence of multiple realizations of the same process. The extension to multiple trials was introduced by [5]. The fitting procedure of the autoregressive (AR) parameters exploits the RLS technique with the use of a forgetting factor. It is based on the minimization of the sum of exponentially weighted prediction errors of the previous processes. Thereby the weighting depends on an adaptation constant $0 \leq C < 1$ that controls the compromise between adaptation speed and the quality of the estimation. Values close to zero result in a slower adaptation with more stable estimations and vice versa. A mean MVAR model was fitted to a set of trials, each one representing the measurement of the same task. A comprehensive description of the algorithms may be found in [5,6].

7.2   THE SIMULATION STUDY

The simulation was designed in order to test the capability of the two methods to follow rapid changes in the cortical connectivity as well as the precision of the estimation performed under different levels of the factors of interest. Test signals simulating cortical average activations in different regions of the cortex were generated in order to fit an imposed coupling scheme involving three cortical areas (shown in Fig. 7.1A). Simulated signals with different levels of signal-to-noise ratio (SNR) and different number of trials have been systematically generated in order to evaluate the influence of these factors on the estimates produced by the two methods.

Signal x_1 was the average activity of a cortical region of interest obtained by a linear inverse estimation procedure from a real EEG recording. The frequency spectrum of x_1 follows the very typical appearance of the EEG signal with a shape resembling the standard $1/f$ function, f being the EEG frequency with a light peak in the alpha band (8–12 Hz). The other signals were generated as

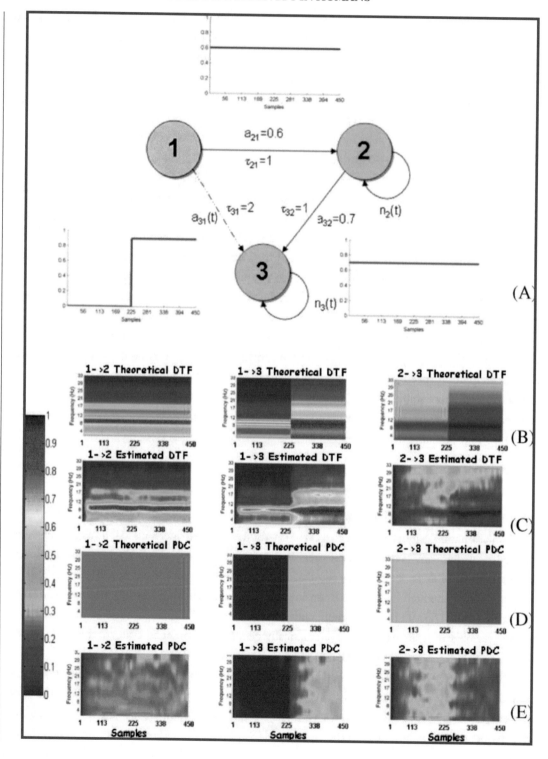

follows:

$$x_j(t) = \sum_{i=1}^{N} a_{ji}(t)x_i(t - \tau_{ji}) + n_j(t) \quad \text{for} \quad j = 2, \ldots, N \tag{7.1}$$

where N is the number of ROIs; τ_{ji} is the constant delay in the propagation from the ith to the jth area expressed in samples; $a_{ji}(t)$ is the time-dependent amplitude of the connection between the ith and the jth area; $n_j(t)$ is the residual representing the part of the jth area activation not depending from other areas, here playing the role of noise simulated with an uncorrelated Gaussian white noise. In particular, the SNR levels were obtained as the ratio between the power of the signals and the power of the noise signal. This was obtained by adjusting the amplitude of the white Gaussian noise superimposed to the data, in order to match the desired level of SNR. Delays applied were of one and two samples, corresponding to delays of 4 and 8 ms, respectively, at a sampling rate of 256 Hz.

All procedures of signal generation were repeated under the following conditions:

SNR factor levels $= [1, 5, \text{and } 10]$

TRIALS factor levels $= [1, 2, 3, 5, 10, 20, 40, 80]$ each of 1000 samples, at a sampling frequency of 256 Hz

Adaptation constant C factor levels $= [0.01, 0.02, 0.03, 0.04, 0.05, 0.06, 0.07, 0.08, 0.09, 0.10]$

The levels chosen for both SNR and TRIALS factors cover the typical range for the cortical activity estimated with high-resolution EEG techniques.

The results obtained by the two estimators were evaluated in four frequency bands, theta (4–7 Hz), alpha (8–12 Hz), beta1 (13–22 Hz), and beta2 (23–30 Hz).

To evaluate both the adaptation speed and the precision of the connectivity patterns estimated by the two methods, two different indices of the performances were defined.

The first index was the time at settling, or t_s, defined as the first instant following the transition after which the error is below the $\varepsilon\%$ of the transition amplitude (in our case, $\varepsilon = 10$), corresponding to the following condition:

$$|\hat{\sigma}_{ij}(t, b) - \sigma_{ij}(t, b)| \leq \varepsilon\Delta_{ij}^k(b)/100, \ \forall t \in k, \quad t > t_s \tag{7.2}$$

FIGURE 7.1: (A) Connectivity pattern imposed between different regions of the cortex during simulation. The values of the connection strengths a_{ij} are time dependent, and are represented by the plots near each connectivity arrow. τ_{ij} represents the constant delay in the propagation from area j to area i, expressed in samples. $\tau_{21} = 1$, $\tau_{31} = 2$ and $\tau_{32} = 1$ (corresponding to 4, 8 and 4 ms, respectively). (B and C) show an example of theoretical and estimated time frequency distribution of the time-varying DTF function for the three significant arcs in the model. (D and E) show the theoretical and the estimated distribution for the time-varying PDC function.

where b represents the generic frequency band, $\sigma_{ij}(t, b)$ and are the theoretical and the estimated average values of the time frequency function (DTF or PDC), in each cortical band b; Δ_{ij}^k is the time average computed during the kth transition interval (in the results here presented, there is only a transition in the simulated data, and $k = 1$) and ε is the percentage of the transition amplitude desired. The values of the DTF or PDC functions analyzed in this study were obtained by averaging the values of DTF or PDC for each frequency belonging to the band considered.

The second index, used to evaluate the precision of the estimation performed, is the average error in each time interval during which the values of connection strengths are kept constant after the transition, and is defined as follows:

$$\overline{E_{ij}^k(b)}^t = \overline{|\hat{\sigma}_{ij}(t, b) - \sigma_{ij}(t, b)|}^t \tag{7.3}$$

Each generation–simulation procedure was repeated 50 times, in order to increase the robustness of the successive statistical analysis.

7.2.1   Statistical Analysis

The results obtained were subjected to separate analysis of variance (ANOVA). The main factors of the ANOVAs were the SNR (three levels: [1, 5, and 10]) the number of TRIALS jointly used for the estimation (eight levels: [1, 2, 3, 5, 10, 20, 40, 80]), the adaptation constant C (10 levels: [0.01, 0.02, 0.03, 0.04, 0.05, 0.06, 0.07, 0.08, 0.09, 0.10], each of 1000 samples, at a sampling frequency of 256 Hz) and the frequency BAND (with four levels: theta = 4–7 Hz; alpha = 8–12 Hz; beta1 = 13–22 Hz; beta2 = 23–30 Hz). Separate ANOVAs for each frequency band were also performed on the error indices defined in the previous paragraph (Relative Error and Time at Settling). In all the evaluated ANOVAs, the correction of Greenhouse–Gasser for the violation of the spherical hypothesis was used. The *post hoc* analysis with Duncan's test at the $p = 0.05$ statistical significance level was then performed.

7.3   RESULTS

Simulated data representing the activations in three cortical areas were generated as described by Eq. (7.1), in order to fit a time-varying connectivity pattern, shown in Fig. 7.1A. The procedure was repeated under all the different conditions of number of trials and SNR. An adaptive MVAR of order $p = 3$ was fitted to each set of simulated data. The model order p was chosen by means of a method proposed in [7]. The multivariate Akaike criterion was applied on each time interval and the maximum order obtained was then adopted for all the recursive estimation.

The estimation of MVAR models with time-varying parameters was repeated with values of the adaptation constant C ranging from 0.01 to 0.1. Time-varying DTF and PDC were then computed, and the ANOVA was performed on the indices of performances described in Eqs. (7.2)

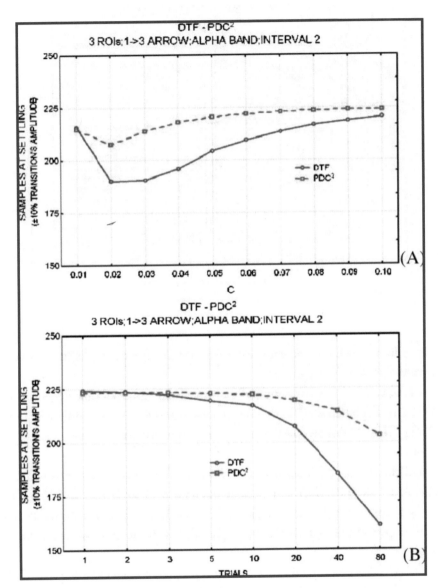

FIGURE 7.2: Results of ANOVA performed on the time at settling (10% of the transition amplitude) for different values of the factors C and TRIALS. (A) Plot of means in function of the adaptation factor C used in the estimation of the AMVAR. A minimum can be noted for values of 0.02 for DTF and 0.02–0.03 for PDC. (B) Effect of different number of trials used for the recursive estimation. A better adaptation speed is shown for higher number of trials. (C) Two-way interaction between factors C and TRIALS. It can be noted that for higher number of trials the minimum shifts to values of C comprised between 0.01 and 0.04. Results are reported in the alpha band (8–12 Hz).

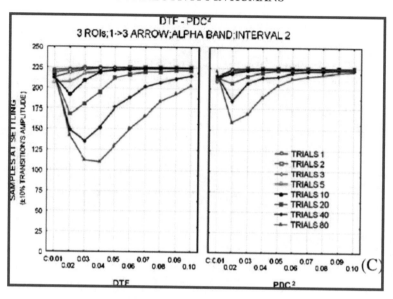

FIGURE 7.2: (*Continued*)

and (7.3), in order to evaluate in a statistically rigorous manner the capability of the method to retrieve the correct connectivity pattern.

Figure 7.1B and C shows an example of theoretical and estimated time frequency distribution of the time-varying DTF function, for the connections imposed to simulated signals ($1 \rightarrow 2$, $1 \rightarrow 3$, and $2 \rightarrow 3$). Fig. 7.1D and E shows the theoretical and the estimated distribution for the time-varying PDC function.

The statistical factors analyzed were the SNR, the number of trials jointly used for the estimation of the AMVAR (TRIALS), and the adaptation factor C. Different ANOVAs addressed the behavior of the indices in the different frequency bands analyzed. However, the results obtained are invariant with respect to the frequency band employed. For this reason, we reported the results in alpha band (8–12 Hz).

The first ANOVA (dependent variable: time at settling) revealed a strong statistical influence of the factors SNR, TRIALS, and C ($p < 0.0001$). In particular, the plots of means for different values of C, TRIALS, and for the two-way interaction between the two factors are shown in Fig. 7.2. It can be noted from Fig. 7.2A that, for both DTF and PDC, there is a minimum in the time at settling for a certain value of C, which is equal to 0.02. This can be explained with a tradeoff between the adaptation speed (which decreases for low values of C) and the variance of the estimation (which decreases with high values of C). Moreover, the adaptation speed of DTF is higher than PDC. From Fig. 7.2B it can be also noted that a high number of trials makes the time at settling decrease.

However, as shown in Fig. 7.2C, it can be seen that an increase in the number of trials makes the minimum of the time at settling shift to higher values of the forgetting factor C. This means that, in practical applications, a higher value of C, and consequently a higher adaptation speed, can be reached if the number of trials at disposal increases.

The second index of performance analyzed was the average error in each time interval during which the values of connection strengths are kept constant after the transition.

ANOVA that was performed revealed a strong influence of the factors SNR, TRIALS, and C in all the frequency bands analyzed. Figure 7.3 shows the effects of different levels of SNR and TRIALS. It can be noted that there is a decrease for higher values of SNR as well as for higher number of trials, for both time-varying DTF and PDC. Moreover, such errors are higher for DTF than for PDC in all conditions. In order to appreciate the comparison with the performances of the classic methods, the results of the application of DTF and PDC based on the estimation of MVAR models are also presented. Figure 7.4 shows the results obtained by the time-varying methods and by the classic MVAR estimation on simulated data with time-dependent connectivity generated according

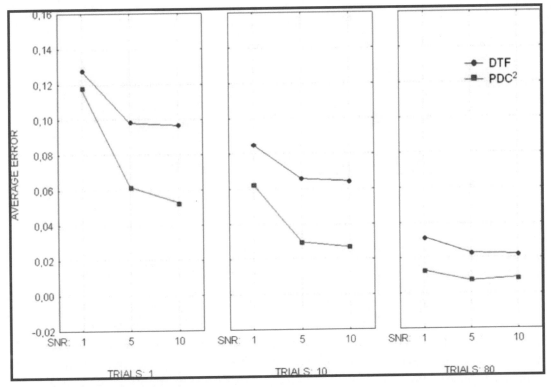

FIGURE 7.3: Results of ANOVA performed on the average error on the arrow $1 \rightarrow 3$ of the model of Fig. 7.1A, in the second time interval (values of connection strengths constant after the transition) on each connectivity arc of the model. $F(4, 196) = 836.44$, $p < 0.00001$.

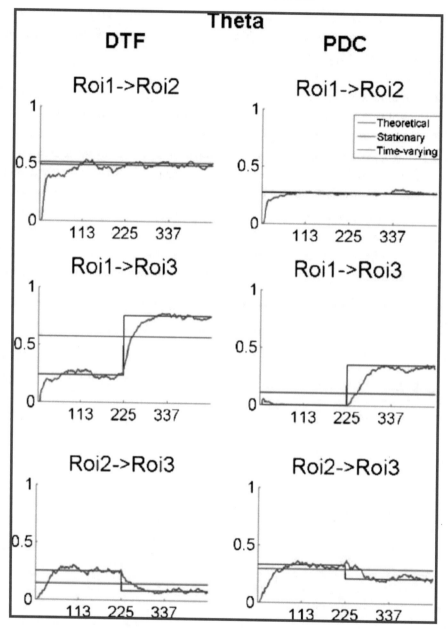

FIGURE 7.4: Comparison of the performances obtained by the time-varying methods and the classic MVAR estimation, on simulated data with time-dependent connectivity. The simulated signals were generated according to the model shown in Fig. 7.1. The time course of the theoretical values of DTF (left column) and PDC (right column) in theta band, for the three connections present in the model (*step line*). The mean value of the results obtained by time-varying DTF and PDC in the same band (*time-varying line*). The mean results obtained by classic DTF and PDC in the same band (*continuous line*).

to the model shown in Fig. 7.1. The time course of the theoretical values of DTF (left column) and PDC (right column), averaged in theta band (4–7 Hz) are shown by the step line. The mean value, in the same band, of the results obtained by time-varying DTF and PDC is shown by the rapid time-varying line. The mean results, in the same band, obtained by classic DTF and PDC are shown by the continuous line. Only the connection imposed in the generation model is shown ($1 \rightarrow 2$, $1 \rightarrow 3$ and $2 \rightarrow 3$).

7.4   DISCUSSION

In this chapter, we propose an evaluation of the performances of two adaptive, multivariate estimators of the cortical connectivity, based on methods usually applied to stationary signals (i.e., PDC and DTF), and to the estimation of an AMVAR with a RLS method. The extension of the RLS algorithm with forgetting factor to MVAR estimation was originally proposed by Moeller et al. [5]. The advantages of this estimation technique are an effective computation algorithm and a high adaptation capability. Simulations on the efficacy of time-variant Granger causality based on AMVAR computed by RLS algorithm were provided previously by [6]. It has been demonstrated [5] that the adaptation capability of the estimation (measured by its adaptation speed and variance) does not depend on the model dimension.

The simulations also address the effect of different factors usually met in the recordings on the performances of the two estimators, and the better choice of the appropriate forgetting factor C under different conditions. The performances of the two estimators were evaluated with particular attention to two aspects: the capability to follow rapid changes in the connectivity patterns (adaptation speed) and the precision in the estimation of connectivity strengths (average error). We performed a series of simulations where test signals were generated to simulate the average electrical activity of cerebral cortical regions, as it can be estimated from high-resolution EEG recordings gathered under different conditions of noise and length of the recordings. Time-varying connectivity was imposed on the simulated signals. The effects of different factors affecting the EEG recordings together with the influence of different choices of the forgetting factor were inferred from statistical analysis (ANOVA and Duncan's *post hoc* tests on the Time at Settling and the Average Error). We also presented the comparison between the performances of the classic and the time-varying formulation of the two estimators, which allows appreciating the improvement obtained with respect to classical methods for non-stationary data.

The results of the simulation study suggested an answer to the questions elicited in the Introduction chapter:

1. The proposed time-varying estimators are effective in retrieving the rapid changes in time of the cortical connectivity pattern. If the operative conditions are properly set, the connectivity pattern can be accurately recovered.

2. The signal-to-noise ratio and the amount of trials at disposal have a statistically significant effect on the performances, both on the side of the adaptation speed and on the side of the accuracy of the estimation. A signal-to-noise ratio of at least 5 is sufficient to obtain a good accuracy, as higher values do not show a significant improvement in the performances. A high number of trials (20 or more) seems to be highly significant, both for improving the accuracy of the estimation and for increasing the adaptation speed.

3. There is an optimum value (between 0.01 and 0.04) for the choice of the adaptation constant C, which varies with the experimental conditions. In particular, a large amount of trials at disposal can increase the value of the optimum choice, thus allowing a higher adaptation speed without loosing the estimation accuracy. In order to obtain a satisfactory adaptation speed, a number of trials not smaller than 20 are required from simulation results.

4. Regarding the comparison between the time-varying DTF and PDC, on the one hand, the DTF method showed better performances of the adaptation speed with lower values of the time at settling. On the other hand, the PDC provided better results of the estimation accuracy with lower values of the average error. The choice between the two methods should take into account the above-mentioned properties, according to the nature of the signals to be examined and of the performances desired, i.e., whether a rapidly changing connectivity is expected or not. It should also be considered that, on the basis of its mathematical formulation, the DTF is unable to discriminate between direct connectivity patterns between areas and indirect links mediated by other areas [3,8–10]. Anyway, for a general purpose, where the interest is in studying the areas that originate the information flow toward the others during a task or a pathological situation, this distinction could not be crucial and DTF could be preferred in order to obtain a higher adaptation speed. Another aspect to be considered is the fact that DTF requires a higher computational effort with respect to PDC, as it involves the inversion of the transfer matrix of the MVAR. So, in order to obtain a faster, more accurate estimation of direct causality links, PDC should be used. This consideration could play a role in the use of these connectivity pattern estimators for on-line applications, such as for instance in the Brain Computer Interface studies.

In conclusion, the statistical results here obtained (by ANOVA integrated with Duncan's *post hoc* tests at $p < 0.05$) indicated an influence of different levels of the main factors SNR, TRIALS, and C on the efficacy of the estimation of time-varying cortical connectivity through time-varying DTF and PDC. In particular, an SNR equal to 5, and at least 20 trials are necessary to decrease significantly the errors on the indices of quality adopted. These conditions are generally obtained in many standard EEG recordings of event-related activity in humans, usually characterized by values of SNR ranging from 3 (movement related potentials) to 10 (sensory evoked potentials) [11].

REFERENCES

[1] M. Kaminski and K. Blinowska, "A new method of the description of the information flow in the brain structures," *Biol. Cybern.*, vol. 65, pp. 203–210, 1991, doi:10.1007/BF00198091.

[2] M. Kaminski, M. Ding, W. A. Truccolo, and S. Bressler, "Evaluating causal relations in neural systems: Granger causality, directed transfer function and statistical assessment of significance," *Biol. Cybern.*, vol. 85, pp. 145–157, 2001, doi:10.1007/s004220000235.

[3] L. A. Baccalà and K. Sameshima, "Partial Directed Coherence: A new concept in neural structure determination," *Biol. Cybern.*, vol. 84, pp. 463–474, 2001, doi:10.1007/PL00007990.

[4] M. Ding, S. L. Bressler, W. Yang, and H. Liang Ding, "Short-window spectral analysis of cortical event-related potentials by adaptive multivariate autoregressive modeling: Data preprocessing, model validation, and variability assessment," *Biol. Cybern.*, vol. 83, pp. 35–45, 2000, doi:10.1007/s004229900137.

[5] E. Möller, B. Schack, M. Arnold, and H. Witte, "Instantaneous multivariate EEG coherence analysis by means of adaptive high-dimensional autoregressive models," *J. Neurosci. Methods*, vol. 105, pp. 143–158, 2001, doi:10.1016/S0165-0270(00)00350-2.

[6] W. Hesse, E. Möller, M. Arnold, and B. Schack, "The use of time-variant EEG Granger causality for inspecting directed interdependencies of neural assemblies," *J. Neurosci. Methods*, vol. 124, pp. 27–44, 2003, doi:10.1016/S0165-0270(02)00366-7.

[7] B. Schack, A. C. N. Chen, S. Mescha, and H. Witte, "Instantaneous EEG coherence analysis during the Stroop task," *Clin. Neurophysiol.*, vol. 110, pp. 1410–1426, 1999, doi:10.1016/S1388-2457(99)00111-X.

[8] M. Winterhalder, B. Schelter, W. Hesse, K. Schwabb, L. Leistritz, D. Klan, R. Bauer, J. Timmer, and H. Witte, "Comparison of linear signal processing techniques to infer directed interactions in multivariate neural systems," *Signal Process.*, vol. 85, no. 11, pp. 2137–2160, Nov 2005, doi:10.1016/j.sigpro.2005.07.011.

[9] R. Kus, M. Kaminski, and K. J. Blinowska, "Determination of EEG activity propagation: Pairwise versus multichannel estimate," *IEEE Trans. Biomed. Eng.*, vol. 51, no. 9, pp. 1501–1510, Sep 2004, doi:10.1109/TBME.2004.827929.

[10] L. Astolfi, F. Cincotti, D. Mattia, M. Lai, F. de Vico Fallani, S. Salinari, L. A. Baccalà, M. Ursino, M. Zavaglia, and F. Babiloni, "Causality estimates among brain cortical areas by partial directed coherence: Simulations and application to real data," *Int. J. Bioelectromagnetism*, vol. 7, no. 1, 2005.

[11] D. Regan, *Human Brain Electrophysiology. Evoked Potentials and Evoked Magnetic Fields in Science and Medicine*, New York: Elsevier, 1989.

CHAPTER 8

Time-Varying Connectivity from Event-Related Potentials

8.1 EXPERIMENTAL DESIGN AND EEG RECORDINGS

Five right-handed healthy subjects (mean age 24.1 ± 1.5) participated in the study after informed consent was obtained. Subjects were seated comfortably in an armchair with both arms relaxed and resting on pillows. They were asked to perform a brisk protrusion of their lips (lip pursing) while they were performing a right foot movement. The task was repeated every 6–7 s, in a self-paced manner, and 100 single trials were recorded. A 58-channel electroencephalography (EEG) system (BrainAmp, Brainproducts GmbH, Germany) was used to record electrical potentials by means of an electrode cap, to an extension of the 10–20 international systems. Structural MRIs of the subject's head were taken with a Siemens 1.5T Vision Magnetom MR system (Germany). The surface electromyographic (EMG) activity of the muscle was also collected. The onset of the EMG response served as zero time. All data were visually inspected, and trials containing artifacts were rejected. A semiautomatic supervised threshold criteria for the rejection of trials contaminated by ocular and EMG artifacts was used, as described in details elsewhere [1]. After the EEG recording, the electrode positions were digitized using a three dimensional (3D) localization device with respect to the anatomic landmarks of the head (nasion and two preauricular points). The analysis period for the potentials time-locked to the movement execution was set from 1500 ms before to zero time (EMG trigger).

8.2 HEAD AND CORTICAL MODELS

The generation of the head and cortical models as well as the estimation of the cortical activity in the particular region of interest (ROIs) of the obtained model has been adequately treated in the previous chapters and will not be repeated here.

8.2.1 Regions of Interest

The particular cortical regions of interest (ROIs) employed in this study were drawn by an expert neuroradiologist on the computer-based cortical reconstruction of the individual head models of five subjects. Twelve ROIs, thought to be involved in the preparation and execution of simple self-generated movements, were defined on the cortical model: left and right supplementary motor area proper (SMAp); the caudal cingulate motor area (CMAc) from the left and right hemispheres; the primary motor foot (MI-foot) representational area and the primary motor lip (MI-lip) representational area, both from the left and right hemisphere; the superior parietal cortex, SP and the premotor dorsal cortex (PMd), both in the left and right hemispheres.

8.3   RESULTS

The estimation of the current density waveforms in the ROIs employed was performed by means of the linear inverse procedure, for five subjects, according to Eqs. (3.1)–(3.4) of Chapter 3. Relevant cortical activity was different from baseline in SMAp, CMAc, MI-foot, MI-lip, SP, and PMd. On such cortical waveforms, connectivity estimations were performed by the time-varying partial direct coherence/direct transfer function (PDC/DTF) procedure as described in the Methods section. The adaptation constant C was set to 0.02, according to the results of the simulation study. The model order was rather stable for all the subjects investigated; for four of them we selected the order 16 and for the last one the order 17.

Figures 8.1 and 8.2 show the time–frequency distribution of the instantaneous PDC and DTF respectively, obtained for a representative subject from the set of cortical waveforms estimated in the 12 ROIs considered. The time interval is relative to 1500 ms preceding the EMG onset, which served as zero time, while the frequency range is between 0 and 50 Hz (see the inset). The figure should be read from column (areas originating the causal influence) toward the rows (target of the causal influence). The first 50 ms reflect the adaptation epoch of the model. Both methods (PDC and DTF) presented comparable results. High values of connectivity can be noted in particular in the alpha and gamma bands, and involved mainly the PMd areas from the left and right hemispheres, the M1F left and right, the SMAp left and right, and the CMAc left and right.

Figure 8.3 shows the time-varying connectivity patterns in the alpha band, extracted at −500, −250 and 0 ms before the movement onset, for the same representative subject of Figs. 8.1 and 8.2. Results are presented on the realistic reconstruction of the head and cortex of the subject, obtained from sequential MRIs. The different ROIs selected are depicted in different grey levels and described by the labels. The connectivity links are represented by arrows, pointing from one cortical area ("the source") toward another one ("the target"). The grey level and size of the arrows code the interaction strength (see color bar on the right of the figure). From these results, it is possible to note some time-invariant patterns as well as some changes during the period of the movement preparation. In particular, it can be observed that the connections between the PMd_Right area and the M1F_Rigth area, between the PMd_Left to the M1Lips_Left and between the SMA_Left and M1F_Left, reveal a strong increase in their strength during the movement preparation, for both DTF and PDC. On the other side, some connectivity links remain constant during the interval. In particular, strong connections are observed from the PMd_R area to the M1Lips_R area, from the CMA_R to the M1F_R and to the SMA_R, for both DTF and PDC and throughout the period examined.

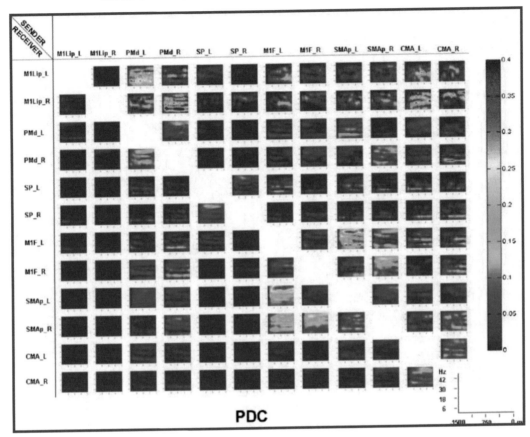

FIGURE 8.1: Time–frequency distribution of the instantaneous PDC obtained for a representative subject, from the set of cortical waveforms estimated in the 12 ROIs considered. On the *x*-axis, the time interval (1500 ms preceding the EMG onset, which served as zero time). On the *y-axis*, the frequencies ranging from 0 to 50 Hz (see the inset at the bottom of the figure). The figure should be read from column (areas originating the causal influence) toward the rows (areas to which the causal influence is directed). The first 50 ms reflect the adaptation epoch of the model. The color bar on the right of the figure represents the intensity of the time-varying connectivity.

Some differences can be detected between the connectivity networks elicited by the two connectivity estimators employed (i.e., DTF and PDC). In particular, it can be observed that DTF showed the existence of a stable connection between the cortical premoter area (PMd) of both hemispheres, as well as a direct link between the right premotor areas (PMd_R) and the left primary motor area for the lips (MILips_L) during all the observed time interval, whereas by using the PDC estimator, these links are not observed at all. In these cases, it could be noted that often the cortical network estimated with the DTF that differs from the PDC networks also admits an alternative path connecting such areas. In particular, the link between the right PMd area and the MILips_L

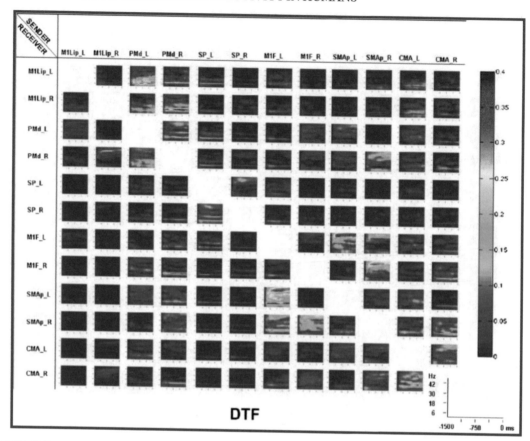

FIGURE 8.2: Time-frequency distribution of the instantaneous DTF obtained for a representative subject, from the set of cortical waveforms estimated in the 12 ROIs considered. Same conventions as in Fig. 8.1.

can also be replaced by a path that starts from PMd_R area, arrives on the SMA_L area and from there toward the PMd_L. In addition, the same path with another step connects the PMd_R with the MILips_L.

Figure 8.4 shows the time course of interaction strength between selected regions of interest from all five subjects analyzed, computed with the PDC estimator. The thin waveforms in different grey levels refer to the connectivity strengths computed for each particular subject of the group. The thick black line is the average time course of connectivity strengths for all the population. Time is expressed in multiseconds from the EMG onset, which served as zero time. The waveforms were normalized to allow comparison between signals with different power spectra, by subtracting their mean values and dividing by their standard deviations. The ROIs presented are the Brodmann area 6 of the left hemisphere (PMd_L), the cingulated motor area left (CMA_L), the primary motor area

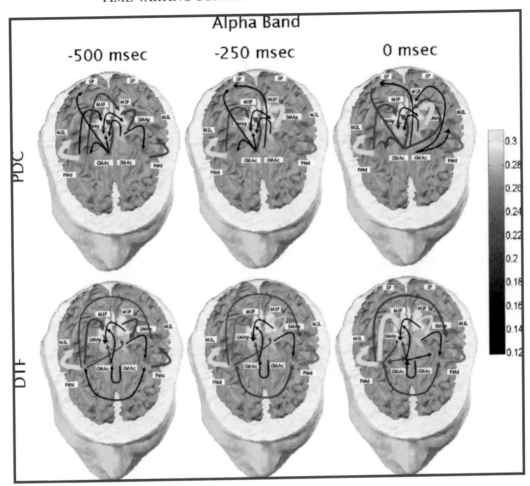

FIGURE 8.3: Time-varying connectivity patterns in the alpha band, extracted at −500, −250 and 0 ms before the movement onset. (*First row*) Results obtained with time-varying DTF. (*Second row*) Results of time-varying PDC. Reconstruction of the head and cortex of the subject, obtained from sequential MRIs. The different ROIs selected are depicted in different grey levels and described by the labels. The grey level and size of the arrows code the interaction strength (see color bar on the right). Similarities can be seen between the results obtained with the two methods as well as an evolution in time of the connectivity during the movement preparation.

for the foot movement, right and left (MIF_R and MIF_L), the primary motor area for the lips, right (A4lip_R). It is possible to recognize a substantial agreement between the subjects across the average connectivity waveform in the framework of the biological variability. In particular, it is of interest that the behavior of the primary motor area of the foot (MIF_R and MIF_L) whose connectivity strengths towards the primary area of the lips (A4Lips) increases starting from 250 ms before the

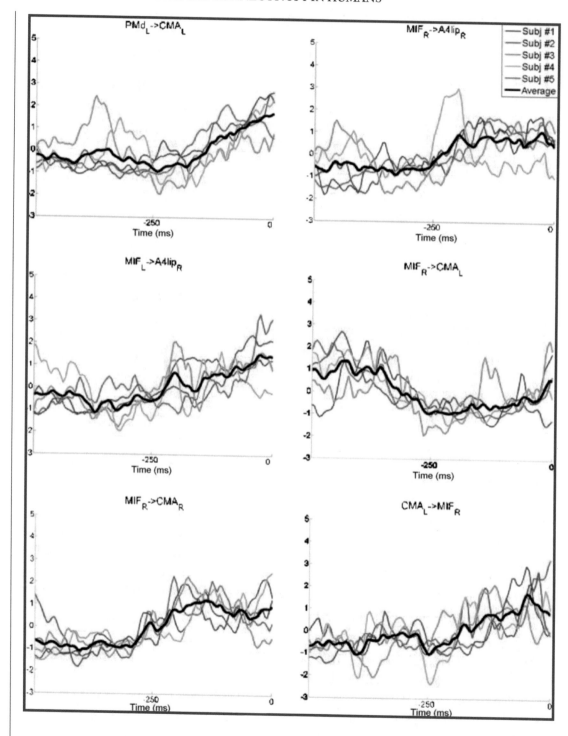

EMG onset. A decrease of connectivity strengths is also observed during the period of time from 500 ms to the 250 ms before the EMG onset between the foot motor area and the CMA.

8.4 DISCUSSION

After being tested by the simulation study, the methods analyzed were applied to the cortical activity estimated in some regions of the cortex of the five subjects under study to highlight the possibilities of the methods of analysis. The information obtained by the simulation study was used to evaluate the applicability of this method to actual event-related recordings. The gathered ERP signals related to the combined foot–lips movement showed an SNR between 3 and 5. The trials at disposal for each experimental subject after the artifact rejection were about 100. According to the simulation results, we applied the time-varying PDC/DTF methods on the estimated cortical current density data expecting a limited amount of errors in the estimation of cortical connectivity patterns.

The problem of the linearity of the coupling between the activities in different areas was also addressed. The evaluation of several methods for the computation of the functional connectivity between couple of EEG/MEG signals was recently performed [2]. It has been concluded that although nonlinear methods, such as mutual information, nonlinear correlation and generalized synchronization [3–5], might be preferred when studying EEG broadband signals that are sensitive to dynamic coupling and nonlinear interactions expressed over many frequencies, the linear measurements are still convenient as they afford a rapid and straightforward characterization of functional connectivity. A recent study [6] applied time-varying DTF and PDC to simulated data with nonlinear coupling (generated by a stochastic Roessler system), demonstrating that the PDC and the DTF are sensitive in detecting interactions in nonlinear multivariate systems, and that high model orders (as those employed in this study) are required to describe the nonlinear system sufficiently with a linear MVAR model.

The simple motor task, the preparation to the execution of a foot-lips movement, was chosen, as it is well known and studied in literature. In particular, the time period represented in Figs. 8.2 and 8.3 is not casual. The literature related to the self-paced movement indicate that at the 500 ms there is on average a brisk increase of the potential negativity over the scalp surface in

←

FIGURE 8.4: Time-courses of connectivity strength between selected ROIs, obtained for the five subjects participating to the study, in the alpha frequency band, by the time varying PDC. Each thin grey waveform refers to the connectivity strengths computed for a particular subject of the analyzed group. The thick black line is the average across subjects of the time-varying connectivity strengths. Time is expressed in milliseconds from the EMG onset (zero time). The waveforms were normalized to allow comparison despite different powers. The ROIs analyzed are the BA 6 of the left hemisphere (PMd_L), the CMA_L, MIF_R and MIF_L, and A4lip_R.

centroparietal regions (last phase of the Bereitschaft potential, BP) and at 250 on average there is a further increase over the central scalp region (negative slope, NS). Furthermore, it has been also shown that at the EMG onset (zero time) the maximum activity is observed on the primary motor area of both hemispheres. Figure 8.4 shows the time-varying PDC connectivity strengths in the population analyzed for a selected couple of ROIs. The average connectivity waveforms (thick black line) showed a consistent pattern of increasing values approaching the EMG onset, suggesting a general increase of connectivity strength between the primary motor areas related to the movement of foot and lips. The observed variability across the population analyzed seems reasonably small, and the same qualitative results were also obtained by computing the time-varying connectivity with the DTF estimator. The physiological results obtained are similar in the five experimental subjects, and seem rather consistent and could integrate those already present in literature on the same movement. The analysis performed with the aid of the time-varying connectivity estimators here derived suggested the existence of two different cortical networks, subserving the preparation and the execution of the joint movement of lip and foot.

In particular, one network is time-invariant during the preparation of the movement involves that PMd, CMA, SMA, M1Lips, and M1F ROIs located on the right hemisphere. It is worth noting that both the connectivity estimators employed (i.e., PDC and DTF) returned the same information. Results obtained suggest that such links connect cortical ROIs that are rather involved in the resource allocation for the behavioral task performed instead of in the monitoring of the actual execution of the joint movement. In this respect, the cortical network depicted by such time-invariant links does not change its activation state during the preparation of the joint movements of lips and foot. In fact, PMd and CMA are ROIs that are activated continuously during the preparation of the movement and generate the sequence of generic activation command towards the primary motor areas responsible for the activation of the specific motorneurons that will trigger the movement for lips and foots.

The second network of cortical areas involved in the preparation and generation of the joint movements is instead dependent on the time-course of the task. In particular, the results suggested the existence of a network that increases in strength during the few hundreds of milliseconds preceding the actual execution of the movement. Such a network involves connections between the PMd_Right area and the M1F_Rigth area, between the PMd_Left to the M1Lips_Left and between the SMA_Left and M1F_Left. Once again, it is worth noting that such increase in connectivity strengths has been detected by both the connectivity estimators applied, i.e., both PDC and DTF. Differences between the cortical networks estimated by the DTF when compared to the PDC are related to the existence of alternative paths between the same cortical areas, which can be modelled by the DTF with also a direct link. This property of the DTF has been established on the basis of its mathematical formulation, as described earlier. However, it must be said that there also exists cortical paths estimated with the PDC that are not estimated with the DTF. For these discrepancies, which are relative to a few connections on the total amount of the estimated cortical networks, possible explanations are relative to the different sensitivities of such algorithms to the amount of noise present on the data. The global conclusion, however, is that the cortical networks estimated by these two methods are rather similar, and allows to conclude positively about the applicability of these estimators on real EEG recordings. It is worth noting that the information on the increasing connectivity strength in the second cortical network would be lost with the application of the standard connectivity estimators, which assumes the stationarity of the analyzed data.

In the last two chapters we presented two multivariate causality estimators to retrieve rapidly changing cortical connectivity. Simulations suggest that the methods were adequate to estimate cortical connectivity under a large range of SNR and TRIALS factors, normally encountered in the standard practice of the high-resolution EEG recordings. An SNR of 5 and a number of trials of at least 20 provide a good accuracy in the estimation. The results pointed out that the DTF estimator can assure a higher adaptation speed than PDC, but PDC ensures a better accuracy in the estimation of connectivity strengths. Possible values for the optimum choice of the adaptation factor C have been proposed, according to the different operative conditions. The results of the application of the two methods to real data in a group of subjects revealed time-varying cortical connections during the execution of the task. In particular, two different cortical networks, one constant across the task and the other evolving during the preparation of the joint movement have been detected with the proposed methodology. It is worth noting that the information on the increasing connectivity strength in the second cortical network would be lost with the application of the standard connectivity estimators, which assumes the stationarity of the analyzed data.

In conclusion, this study opens the way to the application to cortical estimations for studying detailed time–frequency patterns describing the evolution of cortical connectivity.

REFERENCES

[1] D. V. Moretti, F. Babiloni, F. Carducci, F. Cincotti, E. Remondini, P. M. Rossini, S. Salinari, and C. Babiloni, "Computerized processing of EEG-EOG-EMG artifacts for multi-centric studies in EEG oscillations and event-related potentials," *Int. J. Psychophysiol.*, vol. 47, no. 3, pp. 199–216, Mar. 2003, doi:10.1016/S0167-8760(02)00153-8.

[2] O. David, D. Cosmelli, and K. J. Friston, "Evaluation of different measures of functional connectivity using a neural mass model," *NeuroImage*, vol. 21, pp. 659–673, 2004, doi:10.1016/j.neuroimage.2003.10.006.

[3] M. S. Roulston, "Estimating the errors on measured entropy and mutual information," *Physica D*, vol. 125, pp. 285–294, 1999, doi:10.1016/S0167-2789(98)00269-3.

[4] C. J. Stam and B. W. van Dijk, "Synchronization likelihood: an unbiased measure of generalized synchronization in multivariate data sets," *Physica D*, vol. 163, pp. 236–251, 2002, doi:10.1016/S0167-2789(01)00386-4.

[5] C. J. Stam, M. Breakspear, A. M. van Cappellen van Walsum, and B. W. van Dijk, "Nonlinear synchronization in EEG and whole head MEG recordings of healthy subjects," *Hum. Brain Mapp.*, vol. 19, no. 2, pp. 63–78, 2003, doi:10.1002/hbm.10106.

[6] M. Winterhalder, B. Schelter, W. Hesse, K. Schwabb, L. Leistritz, D. Klan, R. Bauer, J. Timmer, and H. Witte, "Comparison of linear signal processing techniques to infer directed interactions in multivariate neural systems," *Signal Process.*, vol. 85, no. 11, pp. 2137–2160, November 2005, doi:10.1016/j.sigpro.2005.07.011.

Author Biography

Laura Astolfi is a researcher at the Department of Human Physiology and Pharmacology of the University of Rome "La Sapienza" and at the Neuro-physiopathology Unit of the Fondazione Santa Lucia, Rome. Her research interests involve the study of the cortical connectivity, the neuroelectrical inverse problem and the multimodal integration of fMRI and HR EEG data.

She received the degree in Electronic Engineering *cum laude* from the University of Rome "La Sapienza" in 2003 and the PhD in Biomedical Engineering from the University of Bologna in 2007.

Fabio Babiloni was born in Rome in 1961. He got the master degree in Electronic Engineering at the University of Rome "La Sapienza" summa cum laude in 1986, and the PhD in Computational Engineering at the Helsinki University of Technology, Helsinki in the 2000 with a dissertation on the multimodal integration of EEG and fMRI. He is currently Associate Professor of Human Physiology at the Faculty of Medicine of the University of Rome "La Sapienza", Rome, Italy. He is author of more that 85 papers on bioengineering and neurophysiological topics on international peer-reviewed scientific journals, and more than 200 contributions to conferences and books chapter. Currents interest are in the field of multimodal integration of EEG, MEG and fMRI data, cortical connectivity estimation and Brain Computer Interface. Prof. Babiloni is currently Associate Editor of four scientific Journals "*Clinical Neurophysiology*", "*International Journal of Bioelectromagnetism*", "*IEEE Trans. On Neural System and Rehabilitation Engineering*", and "*Computational Intelligence and Neuroscience*".